Holt McDougal Mathematics

Course 2
Chapter 2 Resource Book

HOLT McDOUGAL
a division of Houghton Mifflin Harcourt

Printed in the United States of America

ISBN 13: 978-0-55-400725-0
ISBN 10: 0-55-400725-8

1 2 3 4 5 170 12 11 10 09 08

Contents

Description of Contents

Family Involvement Pages

The Chapter Resource Book includes a set of family involvement pages for each section. The family involvement pages consist of the following items:

- **Family Letter**, a two-page letter describing the math that the student will study in each section. A list of vocabulary words from the lessons is included, and some worked-out examples are provided.
- **At-Home Practice**, a one-page worksheet with problems drawn from the content described in the Family Letter. Answers are provided on the page so that the student and his or her family can check this work at home.
- **Family Fun**, a one-page activity sheet that the student and his or her family can work on together.

Practice A, B, and C

There are three practice worksheets for every lesson. All of these reinforce the content of the lesson. Practice B is shown in the Teacher's Edition and is appropriate for the on-level student. It is also available as a workbook (Homework & Practice Workbook)

Practice A is easier than Practice B but still practices the content of the lesson. Practice C is more challenging than Practice B.

Review for Mastery

The Review for Mastery worksheet (one per lesson) provides an alternate way to teach or review the main concepts of the lesson. This worksheet is one or two pages long and is shown in the Teacher's Edition.

Challenge

The Challenge worksheet (one per lesson) enhances critical thinking skills and extends the lesson. This worksheet is shown in the Teacher's Edition.

Problem Solving

The Problem Solving worksheet (one per lesson) provides practice in problem solving and opportunities for real-world applications and for interdisciplinary connections. There are both multiple choice and short response problems. This worksheet is shown in the Teacher's Edition.

Reading Strategies

The Reading Strategies worksheet (one per lesson) provides tools to help the student master math vocabulary or symbols.

Puzzles, Twisters & Teasers

The Puzzles, Twisters & Teasers worksheet (one per lesson) provides fun practice while reinforcing the content of the lesson.

CHAPTER 2

Family Letter

2A Integers

Dear Family,

The student will be learning about integers and how these numbers relate to the coordinate plane. The set of integers includes the set of whole numbers (0, 1, 2, 3, ...) and their opposites (–1, –2, –3, ...).

Number lines will be used to introduce the concept of "opposite" numbers. The **opposite** of a number is the same distance from 0 on a number line as the given number. In mathematics, this is called absolute value. Absolute value is designated by the symbol | |. Absolute value is always positive because it denotes distance from zero, and distance cannot be negative.

$$|7| = 7 \qquad |-7| = 7$$

The student will learn how to add, subtract, multiply, and divide integers by following a few very important, yet simple, guidelines.

Guidelines for Adding Integers	
If the signs are the **same,** find the **sum** of the absolute values of the integers and give that sum the same sign of the integers in the problem.	If the signs are **different,** find the **difference** of the absolute values. Use the sign of the integer with the greater absolute value.
Find the sum. –5 + –7	**Find the sum. –1 + 8**
Think: *Find the sum of 5 and 7. The signs are the same so give the sum the sign of the integers.*	***Think:*** *Find the difference between 1 and 8. Because 8 > 1, use the sign of 8.*
–5 + –7 = –12	**–1 + 8 = 7**

To subtract integers, change the subtraction sign to an addition sign and then add the opposite of what is shown. Then use the guidelines for adding integers.

Subtract.

A. **8 – 3**

8 + (–3) = 5 — Add the opposite of 3. Since 8 > 3, use the sign of 8.

B. **–5 – 4**

–5 + (–4) = –9 — Add the opposite of 4. The signs are the same, so use the sign of the integers.

Vocabulary

These are the math words we are learning:

absolute value the distance a number is from zero on a number line

integers the set of whole numbers and their opposites

opposites numbers that are the same distance from zero on a number line

CHAPTER 2

Family Letter

2A Integers continued

To multiply and divide integers, the student will use the same process for multiplying and dividing whole numbers, except that he or she must determine the sign of the product or quotient.

- If the integers have the same sign, the product or the quotient will **ALWAYS** be positive. $-5(-6) = 30$
- If the integers have different signs, the product or quotient will **ALWAYS** be negative. $-5(6) = -30$

The student will practice solving equations containing integers. such as $y - (-4) = -9$. To solve for y, you must get the variable by itself on one side of the equation. This is called isolating the variable, and it involves using inverse operations on both sides of the equation. Remember that inverse operations "undo" one another.

Solve each equation.

A. $\mathbf{y - (-4) = -9}$

$$y - (-4) = -9$$
$$\underline{+(-4)} = \underline{+(-4)}$$
$$y = -13$$

(-4) is subtracted from y.
To undo the subtraction and get y by itself, add -4 to both sides of the equation.

B. $\mathbf{-184 = 8n}$

$$184 = 8n$$

8 is multiplied by n.

$$\frac{-184}{8} = \frac{8n}{8}$$
$$-23 = n$$

To undo the multiplication and get n by itself, divide both sides of the equation by 8.

Continue to practice using integers with the student. He or she will use these skills throughout all aspects of mathematics.

Sincerely,

Name ______________________ Date ______________ Class ______________

CHAPTER 2

At-Home Practice

2A Integers

Graph each integer and its opposite on a number line.

1. 3 2. 5 3. –1 4. –6

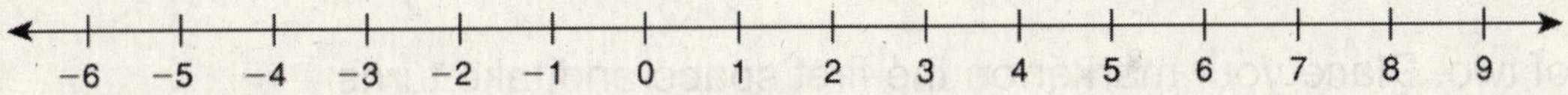

Find each sum.

5. 9 + (–14) 6. –71 + (–63) 7. 25 + (–47)

Find each difference.

8. –42 – (13) 9. –8 – (–7) 10. 9 – 31

Find each product.

11. –8 • 7 12. –4 • (–3) 13. –7 • –3 • 2

Find each quotient.

14. 81 ÷ (–9) 15. –25 ÷ 5 16. –42 ÷ (–6)

Solve. Check your answer.

17. $m + 8 = 4$ 18. $-\frac{1}{6}j = 9$ 19. $-4 = c - 3$ 20. $6d = -24$

Answers: 1–4. Points should be placed on 3, –3, 5, –5, –1, 1, –6, 6. **5.** –5 **6.** –134 **7.** –22 **8.** –55 **9.** –1 **10.** –22 **11.** –56 **12.** 12 **13.** 42 **14.** –9 **15.** –5 **16.** 7 **17.** $m = -4$ **18.** $j = -54$ **19.** $c = -1$ **20.** $d = -4$

Name ______________________ Date ______________ Class ______________

CHAPTER 2

Family Fun

Integer Race

Materials
2 game pieces
Number cube

Play in teams of two. Place your marker on the first space and take turns rolling a number cube. Substitute the number you rolled for the variable in the problem and solve. If the answer is positive, move forward that number of spaces. If the answer is negative, either move back that number of spaces or return to start if that's as far back as you can move. If the answer is zero, stay in that space and lose your turn. First player to the end wins!

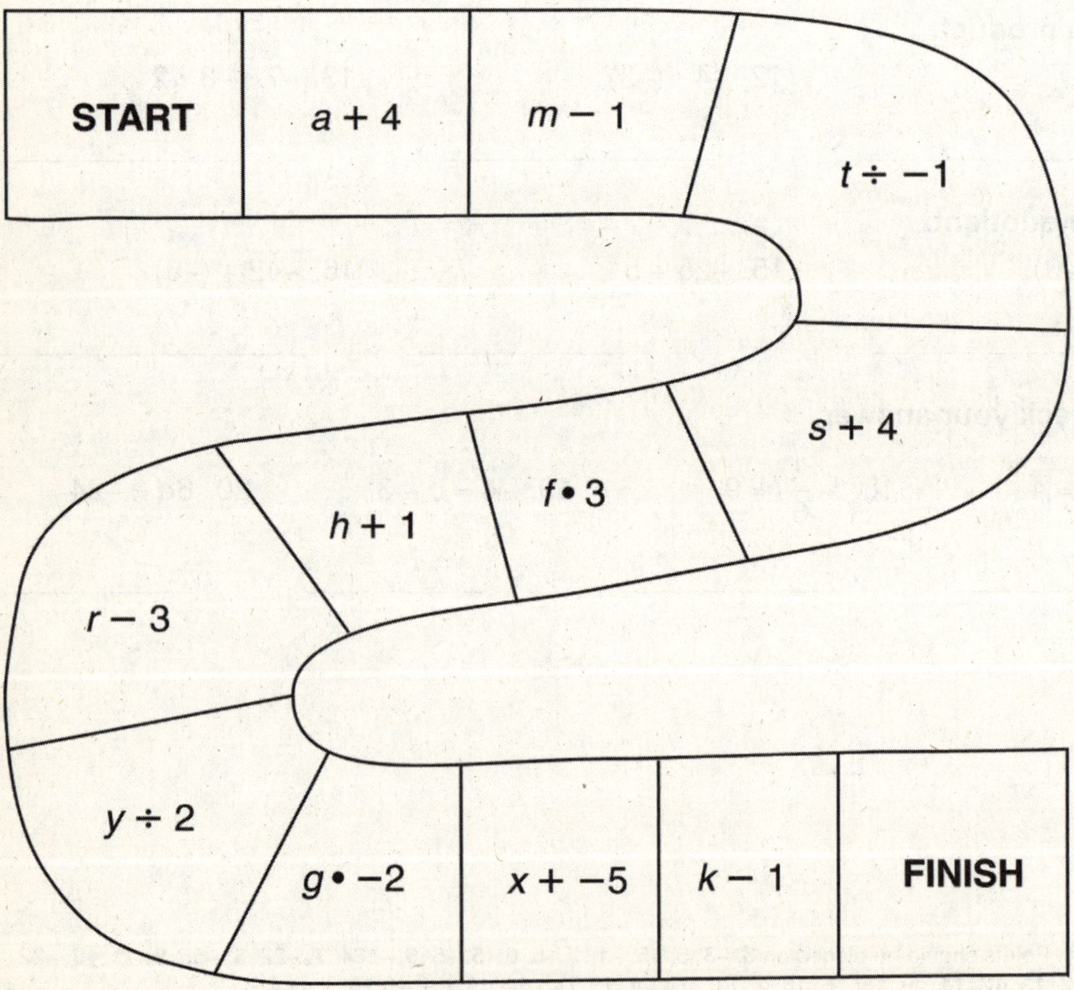

Name ______________________ Date ______________ Class ______________

LESSON 2-1

Practice A

Integers

Graph each integer and its opposite on a number line.

1. 3

−5 −4 −3 −2 −1 0 1 2 3 4 5

2. −5

−5 −4 −3 −2 −1 0 1 2 3 4 5

Use the number line to compare the integers. Write < or >.

−20 −18 −16 −14 −12 −10 −8 −6 −4 −2 0 2 4 6 8 10 12 14 16 18 20

3. −8 ___ 7

4. 4 ___ −7

5. −6 ___ −16

6. −11 ___ 11

Graph the integers on a number line. Then write them in order from least to greatest.

7. −6; 3; −5; 8

−8 −6 −4 −2 0 2 4 6 8

____

8. 6; −7; −8; 0

−8 −6 −4 −2 0 2 4 6 8

____

Use a number line to find each absolute value.

−20 −18 −16 −14 −12 −10 −8 −6 −4 −2 0 2 4 6 8 10 12 14 16 18 20

9. $|-6|$ ______

10. $|2|$ ______

11. $|-1|$ ______

12. $|8|$ ______

13. $|-9|$ ______

14. $|3|$ ______

15. $|-4|$ ______

16. $|10|$ ______

17. $|-15|$ ______

18. $|20|$ ______

19. $|-13|$ ______

20. $|17|$ ______

21. The windchill on a cold day made it feel like 5 degrees below zero outside. Write this temperature as an integer.

____

22. A baby gained 15 pounds from birth to his first birthday. Write this amount as an integer.

____

Name ______________________ Date ______________ Class ______________

LESSON 2-1

Practice B

Integers

Graph each integer and its opposite on a number line.

1. 8

−8 −6 −4 −2 0 2 4 6 8

2. −7

−8 −6 −4 −2 0 2 4 6 8

Compare the integers. Use < or >.

3. −15 ___ −7 4. 8 ___ −8 5. −14 ___ 13 6. −18 ___ −20

Use a number line to order the integers from least to greatest.

7. −1; 4; −5; 7; −3

−8 −6 −4 −2 0 2 4 6 8

____

8. −6; 8; 0; 4; −2

−8 −6 −4 −2 0 2 4 6 8

____

9. 6; 5; −7; −8; −2

−8 −6 −4 −2 0 2 4 6 8

____

10. 1; 3; −4; −5; 7

−8 −6 −4 −2 0 2 4 6 8

____

Use a number line to find each absolute value.

11. $|-18|$ ______ 12. $|11|$ ______ 13. $|-25|$ ______ 14. $|19|$ ______

15. $|-10|$ ______ 16. $|16|$ ______ 17. $|22|$ ______ 18. $|-14|$ ______

19. $|9|$ ______ 20. $|-24|$ ______ 21. $|-7|$ ______ 22. $|17|$ ______

23. Christy dove to a depth of 12 feet below the surface of the water. Write the depth as an integer.

____

24. The highest point in North Carolina is Mt. Mitchell, with a height of 6,684 feet. Write the height of Mt. Mitchell as an integer.

____

Name ______________________ Date ______________ Class ______________

LESSON 2-1

Practice C

Integers

Graph each integer and its opposite on a number line.

1. 15

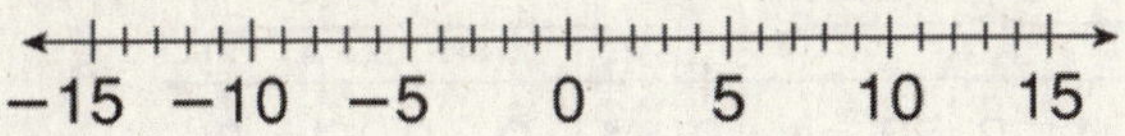

2. –4

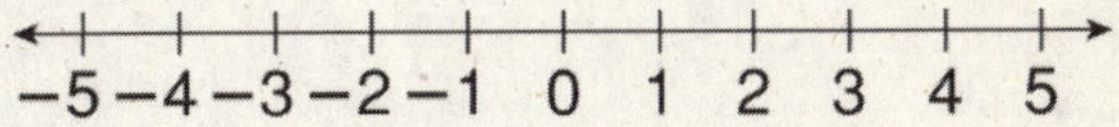

Compare the integers. Use < or >.

3. –15 ___ 15 4. –7 ___ –9 5. 12 ___ –17 6. –13 ___ 9

Use a number line to order the integers from least to greatest.

7. –2; 3; –6; –1; 4

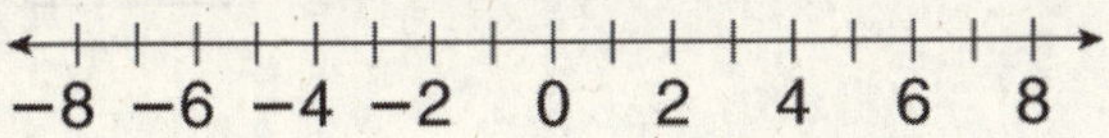

____

8. 7; –5; 6; 0; –7

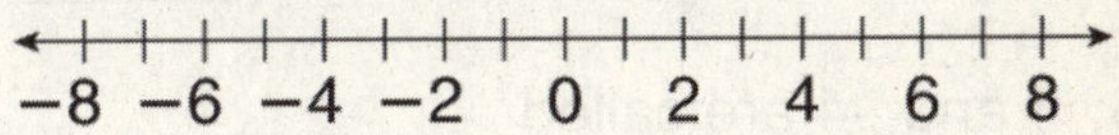

____

Find each absolute value.

9. $|-45|$ ______ 10. $|38|$ ______ 11. $|-99|$ ______ 12. $|22|$ ______

13. $|-19|$ ______ 14. $|375|$ ______ 15. $|-59|$ ______ 16. $|-84|$ ______

Compare. Write <, >, or =.

17. $|-50|$ ___ 50 18. 28 ___ $|-29|$ 19. –43 ___ $|-37|$ 20. $|48|$ ___ 53

21. $|-79|$ ___ $|65|$ 22. –92 ___ $|-92|$ 23. $|-83|$ ___ 82 24. $|34|$ ___ $|-62|$

25. A company reported a net gain of $750,000 for the first quarter and a net loss of $2,000,000 for the second quarter. Write each amount as an integer.

____

26. The record low temperature in Virginia through 1997 was 30°F below 0. The record high temperature was 110°F. Write each temperature as an integer.

____

Name ______________________ Date ______________ Class ______________

LESSON 2-1

Review for Mastery

Integers

This number line shows integers.

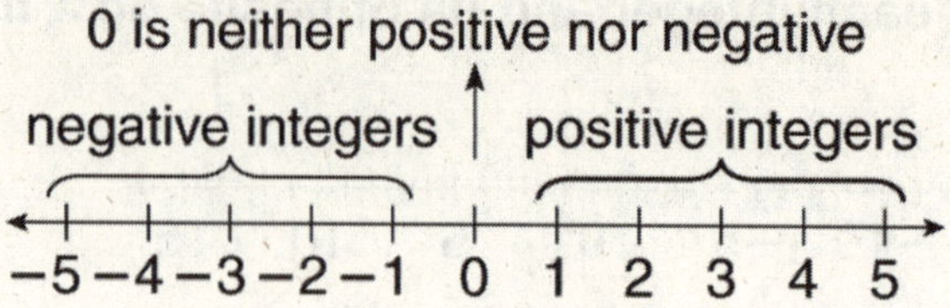

Every integer has an **opposite** integer. A number and its opposite are the same distance from 0.

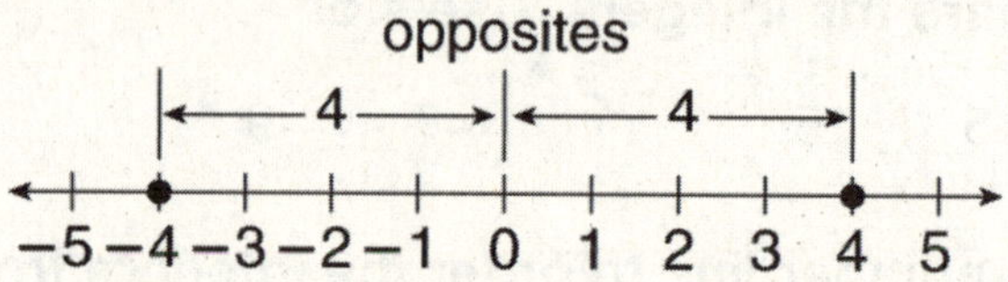

1. How many units is 4 from 0? ________

2. How many units is –4 from 0? ________

3. 4 and –4 are called ______________________________.

Graph each integer and its opposite on a number line.

4. 2

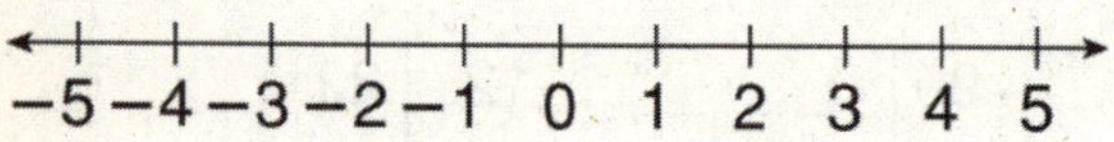

5. –3

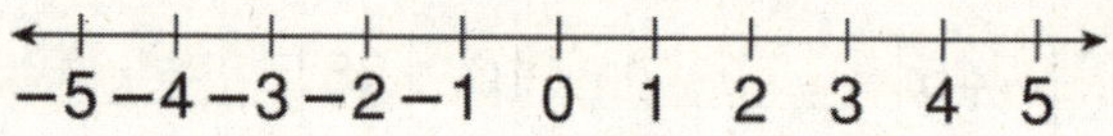

You can use a number line to compare and order numbers. The numbers get greater as you move to the right on the number line.

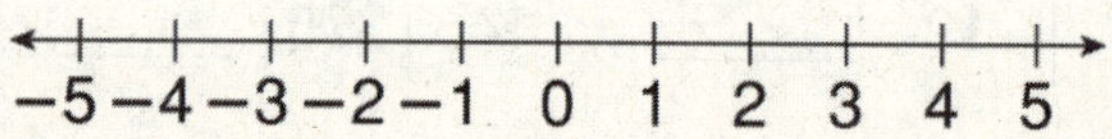

6. What is the order from least to greatest of –1, 2, and –3? ______________

Write the integers in order from least to greatest.

7. –2; –6; 4

8. –3; 7; 1

____

The absolute value of an integer is its distance from 0 on a number line. –5 is 5 units from 0. The absolute value of –5 is 5. You write $|-5| = 5$.

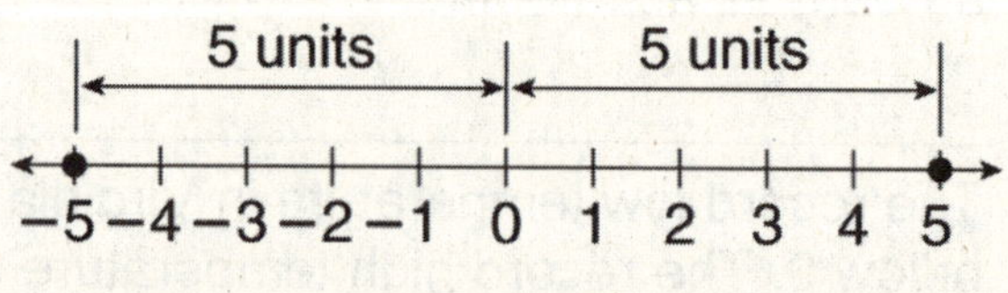

9. How many units from 0 is –3? ______________________________

Name ______________________ Date ______________ Class ______________

LESSON 2-1

Challenge

Order That Integer!

Order each set of integers. Then use the decoder to answer the riddle below.

1. –12; 6; –11

$\_\_\_\_ < \_\_\_\_ < \_\_\_\_$

A E

2. –3; –1; –201

$\_\_\_\_ < \_\_\_\_ < \_\_\_\_$

V S

3. $|-3|$; $|-6|$; –6

$\_\_\_\_ < \_\_\_\_ < \_\_\_\_$

B

4. $|-38|$; $|3|$; –7

$\_\_\_\_ < \_\_\_\_ < \_\_\_\_$

L

5. –4; –40; 4

$\_\_\_\_ > \_\_\_\_ > \_\_\_\_$

O A

6. 0; –8; $|-18|$

$\_\_\_\_ > \_\_\_\_ > \_\_\_\_$

T

7. $|-75|$; $|12|$; 7

$\_\_\_\_ > \_\_\_\_ > \_\_\_\_$

H

8. –16; –12; 1

$\_\_\_\_ > \_\_\_\_ > \_\_\_\_$

U N

What did the car dealer say to the customer?

THIS DEAL IS

____ ____   ____ ____ ____ ____ ____ ____ ____ ____

–4 –12   –4 –6 –1 4 –7 1 0 6

____ ____ ____ ____ ____ .

–3 –4 –7 1 6

Name ______________________ Date ______________ Class ______________

LESSON 2-1

Problem Solving

Integers

Write the correct answer.

1. The coldest place on record in the United States was in Alaska in 1971. It was 80 °F below zero. Write this temperature as an integer.

____

2. The temperature outside was –4 °F at Jared's house and –8 °F at Mario's house. Where was the temperature warmer?

____

3. A small business reported a net loss of $62,500 during its first year. In its second year, it reported a profit of $34,100. Write each amount as an integer.

____

4. For one day, Lacy recorded the low temperatures in five U.S. cities. The temperatures were 5 °C, –1° C, –3 °C, 2 °C, and 0 °C. Write the temperatures in order from least to greatest.

____

Choose the letter for the best answer.

5. Which number is not an integer?

   $-3; 5; \frac{1}{5}; 0$

   A –3 C $\frac{1}{5}$

   B 5 D 0

6. Basha says $|5|$ and $|-5|$ are the same number. Danny says $|5|$ and $|-5|$ are different numbers. Kim says $|5|$ and $|-5|$ both equal 0. Who is correct?

   F Basha

   G Danny

   H Kim

   J They are all wrong.

7. Use the table at right. Which continent has the highest point?

   A Asia C Africa

   B South America D Australia

8. Use the table at right. Which continent has the lowest point?

   F Europe G Australia

   H North America J Asia

Continent	Highest Point (ft)	Lowest Point (ft)
North America	20,320	–282
South America	22,834	–131
Africa	19,340	–512
Asia	29,028	–1,339
Australia	7,310	–52
Europe	18,510	–92

Name ______________________ Date ______________ Class ______________

LESSON 2-1

Reading Strategies

Use a Graphic Organizer

Definition	Facts
The set of whole numbers and their opposites	• Each number can be paired with its opposite. The opposite of 2 is –2. The opposite of –3 is 3. • Zero is its own opposite.
Integers	
Examples	**Non-examples**
0, 2, 5, 9, 13, –3, –7, –12, –17	$\frac{2}{3}$, $\frac{11}{5}$, $2\frac{5}{8}$, 0.5, 0.23, 1.05, 3.61

Answer each question.

1. What are integers?

____

2. Write the opposite of 6. ____________
3. Write the opposite of 10. ____________
4. Write the opposite of 0. ____________
5. Write the opposite of –8. ____________
6. Write the opposite of –3. ____________

Write "integer" or "not an integer" for the following numbers.

7. –9 ____________________
8. $\frac{5}{7}$ ____________________
9. 0.1 ____________________
10. 42 ____________________

Name ____________________ Date ______________ Class ______________

LESSON 2-1

Puzzles, Twisters & Teasers

Food for Thought!

Circle words from the list in the word search. Then find an unlisted word in the word search that answers the riddle. Write the word on the lines below.

negative	opposite	number	integer	absolute
value	set	express	whole	zero

```
M H T E F V C D E S W I C F R
O P P O S I T E V F R N B H U
X Z O I E S W A X N M T A W H
N E G A T I V E C P U E I H Q
U R A S D F G H V L R G P O K
M O Z X C V B N A R J E L L Y
B P O I U W H O L E K R S E M
E D F G H J K L U W E R T S X
R A B S O L U T E Z A W Q E C
```

What kind of fish goes best with peanut butter?

____ ____ ____ ____ ____

Name ______________________ Date ______________ Class ______________

LESSON 2-2

Practice A

Adding Integers

Show the addition on the number line. Then write the sum.

1. 2 + (–3)

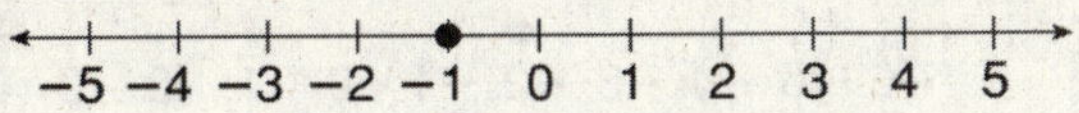

____

2. –3 + (–4)

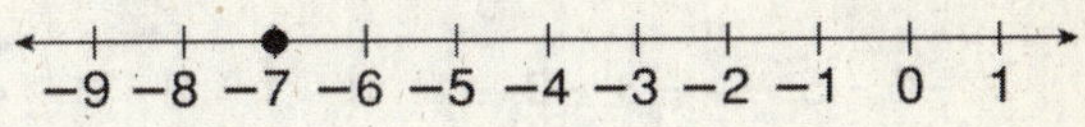

____

Find each sum.

3. –4 + (–9) ________
4. 7 + (–8) ________
5. –2 + 1 ________
6. 6 + (–9) ________
7. 5 + 7 ________
8. 9 + (–5) ________
9. (–1) + 9 ________
10. –9 + (–7) ________
11. 2 + (–7) ________
12. –6 + (–4) ________
13. 3 + 2 ________
14. –2 + 6 ________

Evaluate e + *f* for the given values.

15. *e* = 9, *f* = –2 ________
16. *e* = –4, *f* = –6 ________
17. *e* = 6, *f* = –1 ________
18. *e* = –3, *f* = 2 ________
19. *e* = 8, *f* = –6 ________
20. *e* = –2, *f* = –3 ________

21. The temperature dropped 13 °F in 7 hours. The final temperature was –2 °F. What was the starting temperature?

____

22. A football team gains 8 yards in one play, then loses 5 yards in the next. What is the team's total yardage?

____

Name ______________________ Date ______________ Class ______________

LESSON 2-2

Practice B

Adding Integers

Use a number line to find each sum.

1. –1 + 5

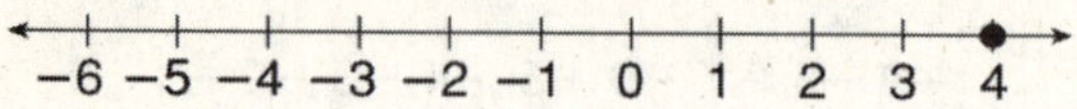

____

2. 4 + (–6)

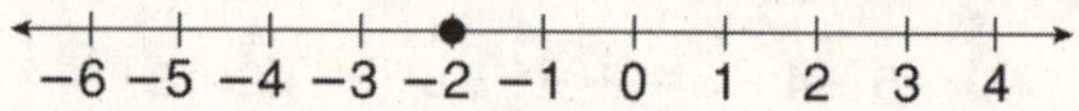

____

Find each sum.

3. –51 + (–9) ________

4. 27 + (–6) ________

5. 1 + (–30) ________

6. 15 + (–25) ________

7. 50 + (–7) ________

8. –19 + (–15) ________

9. (–23) + 9 ________

10. –19 + (–21) ________

11. –17 + 11 ________

12. 20 + (–8) ________

13. (–15) + (–7) ________

14. 12 + (–14) ________

Evaluate $e + f$ for the given values.

15. $e = 9, f = -24$ ________

16. $e = -17, f = -7$ ________

17. $e = 32, f = -19$ ________

18. $e = -15, f = -15$ ________

19. $e = -20, f = 20$ ________

20. $e = -30, f = 12$ ________

21. The temperature rose 9 °F in 3 hours. If the starting temperature was –5 °F, what was the final temperature?

____

22. Matt is playing a game. He gains 7 points, loses 10 points, gains 2 points, and then loses 8 points. What is his final score?

____

Name ______________________ Date ______________ Class ______________

LESSON 2-2

Practice C

Adding Integers

Use a number line to find each sum.

1. $-7 + 10$

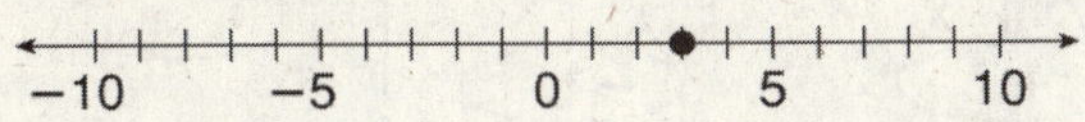

____

2. $-2 + (-6)$

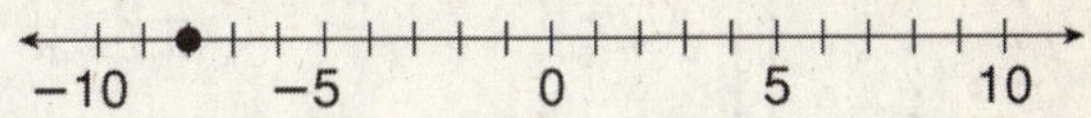

____

Find each sum.

3. $-45 + 9$ ________ 4. $37 + (-60)$ ________ 5. $-11 + 40$ ________ 6. $95 + (-25)$ ________

7. $56 + (-37)$ ________ 8. $-79 + (-15)$ ________ 9. $(-63) + 49$ ________ 10. $-49 + (-47)$ ________

Evaluate each expression for $a = -15$, $b = -8$, and $c = 6$.

11. $14 + b$ ____ 12. $c + (-8)$ ____ 13. $a + b$ ____ 14. $c + 5 + a$ ____

Evaluate $e + f$ for the given values.

15. $e = 90$, $f = -52$ ________ 16. $e = -57$, $f = -23$ ________ 17. $e = -61$, $f = 32$ ________

Compare. Write <, >, or =.

18. $25 + (-25)$ ____ 0 19. $32 + (-19)$ ____ -51 20. $24 + (-42)$ ____ 66

21. The temperature dropped 10 °F in 5 hours. If the starting temperature was 9 °F, what was the final temperature?

____

22. The closing price of a stock gained 2 points on Monday, lost 5 points on Tuesday, lost 1 point on Wednesday, gained 4 points on Thursday, and lost 6 points on Friday. What was the net gain or loss for the week?

____

Name ______________________ Date ______________ Class ______________

LESSON 2-2

Review for Mastery

Adding Integers

This balance scale "weighs" positive and negative numbers. Negative numbers go on the left of the balance, and positive numbers go on the right.

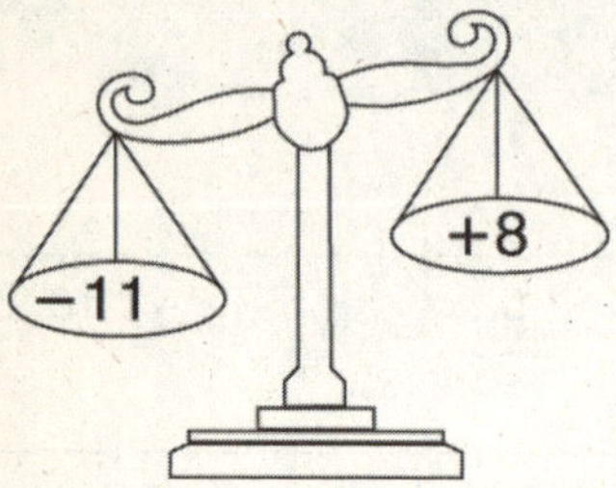

Find −11 + 8.

The scale will tip to the left side because the sum of −11 and +8 is negative.

−11 + 8 = −3

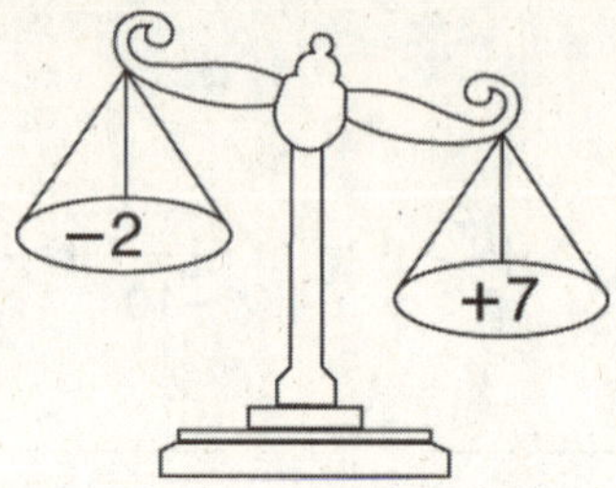

Find −2 + 7.

The scale will tip to the right side because the sum of −2 and +7 is positive.

−2 + 7 = 5

Find −1 + (−3).

Both −1 and −3 go on the left side. The scale will tip to the left side because the sum of −1 and −3 is negative.

−1 + (−3) = −4

Find 3 + (−9).

1. Should you add or subtract? ______________________
2. Will the sum be positive or negative? ______________________

3 + (−9) = −6

(the sign of the integer with the greatest absolute value) (|9| − |3|)

Find −5 + (−8).

3. Should you add or subtract? ______________________
4. Will the sum be positive or negative? ______________________
5. −5 + (−8) = ______________________

Add.

6. 7 + −3 = _____ 7. −2 + −3 = _____ 8. −5 + 4 = _____

9. −3 + −1 = _____ 10. −7 + 9 = _____ 11. 4 + −9 = _____

Name ______________________ Date ______________ Class ______________

LESSON 2-2

Challenge

Alphabet Addition

Find the value of each word. Each vowel has a value as shown in the table. All consonants have a value of 2.

A	E	I	O	U
–5	–11	–8	–3	–6

1. ALGEBRA ____________
2. INTEGER ____________
3. POSITIVE ____________
4. NEGATIVE ____________
5. SIGN ____________
6. ADDITION ____________
7. GREATER ____________
8. LESS ____________
9. EQUAL ____________

Write <, >, or = to compare the values of each word pair.

10. MANY __ FEW
11. ALL __ NOTHING
12. SCHOOL __ HOME
13. DOG __ CAT
14. ADD __ SUBTRACT
15. STOP __ GO
16. EVALUATE __ SOLVE
17. MORE __ LESS
18. EMPTY __ FULL
19. Write your name and the first names of 3 other friends or family members. Find the value of each name. Write the names in order from least to greatest value.

____

20. Write a word with a value of 3.

____

21. Write a word with a value of –3.

____

22. What is the highest value word you can think of? Compare it with your classmates' words.

____

Name ______________________ Date ______________ Class ______________

LESSON 2-2

Problem Solving

Adding Integers

Write the correct answer.

1. The temperature dropped 12 °F in 8 hours. If the final temperature was –7 °F, what was the starting temperature?

____

2. At 3 P.M., the temperature was 9 °F. By 11 P.M., it had dropped 31 °F. What was the temperature at 11 P.M.?

____

3. Tad owes John $23 and borrows $12 more. How much does Tad owe John now?

____

4. New Orleans, Louisiana, is 6 feet below sea level. The highest point in Louisiana, Driskill Mountain, is 541 feet higher than New Orleans. How high is Driskill Mountain?

____

5. A submarine submerged at a depth of –40 ft dives 57 ft more. What is the new depth of the submarine?

____

6. An airplane at 20,000 ft drops 2,500 ft in altitude. What is the new altitude?

____

Choose the letter for the best answer.

7. Last week, Jane made deposits of $64, $25, and $37 into her checking account. She then wrote checks for $52 and $49. What is the overall change in Jane's account balance?

   A –$99 C $126
   B $25 D $227

8. In Indianapolis, Indiana, the coldest recorded temperature was –23 °F. The hottest recorded temperature was 127 °F higher. What was the hottest temperature in Indianapolis?

   F 150 °F H 104 °F
   G 127 °F J –150 °F

9. Helena borrowed $189 from her parents to buy an electric bass. She paid back $56 last week and $64 this week. How much does Helena still owe her parents?

   A $133 C $69
   B $120 D $29

10. The Aral Sea and the Caspian Sea are actually lakes. The elevation of the Caspian Sea is 92 feet below sea level. The Aral Sea is 217 feet higher. What is the elevation of the Aral Sea?

   F –125 ft H 309 ft
   G –309 ft J 125 ft

Name ______________________ Date ______________ Class ______________

LESSON 2-2

Reading Strategies

Use Graphic Aids

Randy's football team had the ball on the zero yard line. On their first play they gained six yards. On the second play they lost four yards. On what yard line is the ball now?

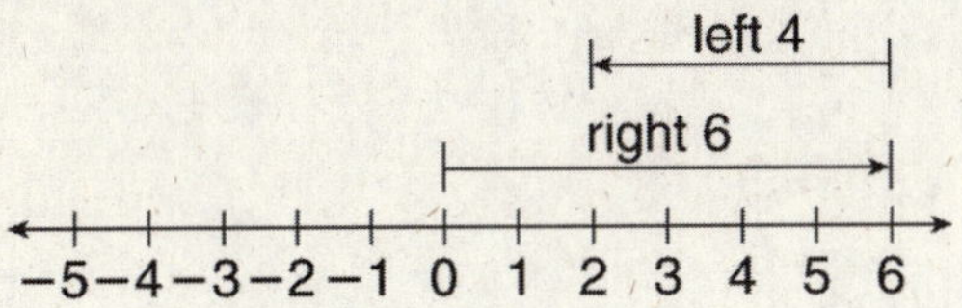

$6 + (-4) = 2$

Use the number line to help you answer the questions.

1. On which number do you begin? ____________

2. Which direction do you move first? How many places do you move?

____

3. Which direction do you move next? How many places do you move?

____

When Angela went to bed, the temperature was zero degrees. When her mother went to bed two hours later, the temperature had gone down 5 degrees. By the time Angela got up the temperature had gone down another 3 degrees. What was the temperature when she got up?

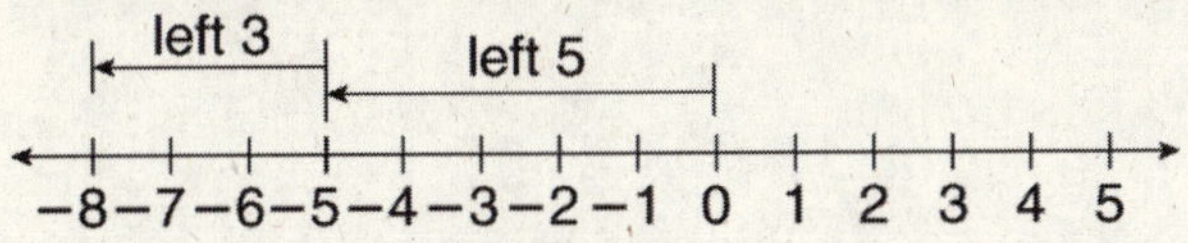

$-5 + (-3) = -8$

Use the number line to help you answer the questions.

4. On which number do you begin? ____________

5. Which direction do you move first? How many spaces?

____

6. Which direction do you move next? How many spaces?

____

Name ______________________ Date ______________ Class ______________

LESSON 2-2

Puzzles, Twisters & Teasers

Chill Out!

Add each pair of integers. Write the letter on the line above the correct answer at the bottom of the page to solve this riddle:

Why did the girl put her money in the freezer?

R 7 + (–16)

S –3 + 10

L –6 + (–13)

H –7 + (–81)

O –1 + (–5)

E 2 + (–20)

C 25 + 14

W –3 + 28

D 9 + 4

A –5 + (–12)

T 6 + (–22)

N 8 + (–7)

___ ___ ___   ___ ___ ___ ___ ___ ___

7 –88 –18   25 –17 1 –16 –18 13

___ ___ ___ ___   ___ ___ ___ ___   ___ ___ ___ ___!

39 –6 –19 13   –88 –17 –9 13   39 –17 7 –88

Name ____________ Date ____________ Class ____________

LESSON 2-3

Practice A

Subtracting Integers

Show the subtraction on the number line. Then write the difference.

1. 3 – 8

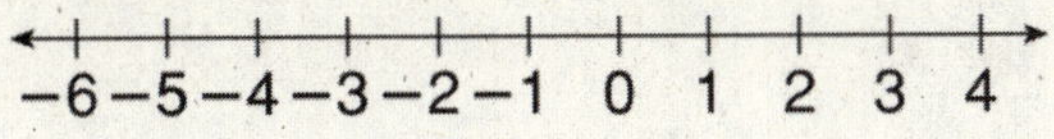

2. –5 – (–1)

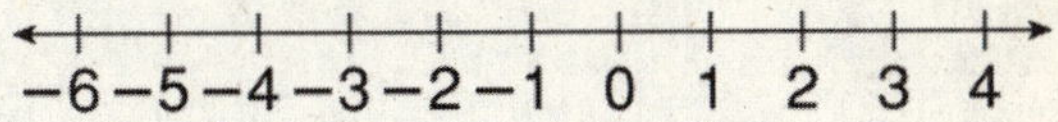

Find each difference.

3. –3 – 4
4. –7 – (–2)
5. 12 – 6
6. 2 – (–7)
7. –8 – 8
8. –5 – (–5)
9. –1 – (–2)
10. 9 – (–3)
11. 8 – 1
12. 7 – (–9)
13. –3 – 8
14. –3 – (–7)

Evaluate *x* – *y* for each set of values.

15. $x = 6$, $y = -3$
16. $x = -7$, $y = 1$
17. $x = -2$, $y = -5$
18. $x = 9$, $y = 11$
19. $x = -1$, $y = -1$
20. $x = -5$, $y = 5$

21. The high temperature one day was 6 °F. The low temperature was –3 °F. What was the difference between the high and low temperatures for the day?

22. The temperature changed from –7 °F at 6 A.M. to 7 °F at noon. How much did the temperature increase?

Name ______________________ Date ______________ Class ______________

LESSON 2-3

Practice B

Subtracting Integers

Use a number line to find each difference.

1. –2 – 3

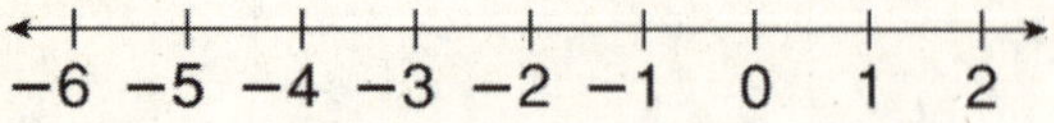

2. 5 – (–1)

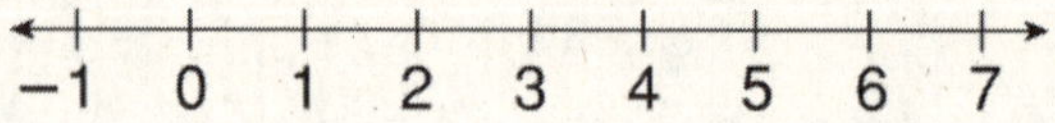

Find each difference.

3. –6 – 4
4. –7 – (–12)
5. 12 – 16
6. 5 – (–19)
7. –18 – (–18)
8. 23 – (–23)
9. –10 – (–9)
10. 29 – (–13)
11. 9 – 15
12. –12 – 14
13. 22 – (–8)
14. –16 – (–11)

Evaluate *x* – *y* for each set of values.

15. $x = 14$, $y = -2$
16. $x = -11$, $y = 11$
17. $x = -8$, $y = -15$
18. $x = -9$, $y = -9$
19. $x = 9$, $y = -20$
20. $x = 20$, $y = 25$

21. The high temperature one day was –1 °F. The low temperature was –5 °F. What was the difference between the high and low temperatures for the day?

22. The temperature changed from 5 °F at 6 P.M. to –2 °F at midnight. How much did the temperature decrease?

Name ______________________ Date ______________ Class ______________

LESSON 2-3

Practice C

Subtracting Integers

Use a number line to find each difference.

1. –6 – (–3)

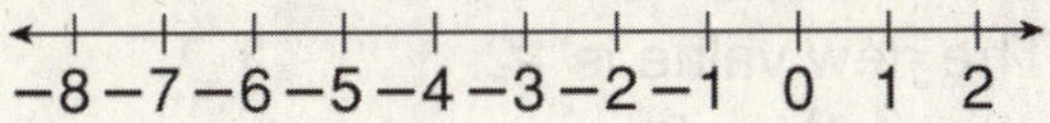

____

2. –1 – 5

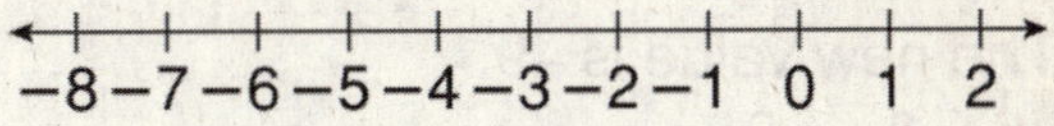

____

Find each difference.

3. –16 – 15 ______________

4. –24 – (–12) ______________

5. 25 – 41 ______________

6. 19 – (–19) ______________

7. 48 – (–18) ______________

8. –37 – 37 ______________

9. –50 – (–50) ______________

10. –39 – 13 ______________

Evaluate $x - y$ for each set of values.

11. $x = -24$, $y = 24$ ______________

12. $x = -22$, $y = -32$ ______________

13. $x = 18$, $y = 25$ ______________

Simplify.

14. 9 + (–15) – 6 ______________

15. 12 – (–5) – 3 ______________

16. –13 – (–8) – (–5) ______________

17. –4 – 9 + (–7) ______________

18. 10 + (–3) – (–18) ______________

19. –25 – 37 – (–5) ______________

20. The record high temperature in North Carolina is 110 °F. The record low temperature is –34 °F. What is the difference in these temperatures?

____

21. The temperature at 6 P.M. was 15 °F. By 8 P.M. it had dropped 9 degrees, and then by 10 P.M. it had dropped another 11 degrees. What was the temperature at 10 P.M.?

____

Name ______________________ Date ______________ Class ______________

LESSON 2-3

Review for Mastery

Subtracting Integers

The total value of the three cards shown is –6.

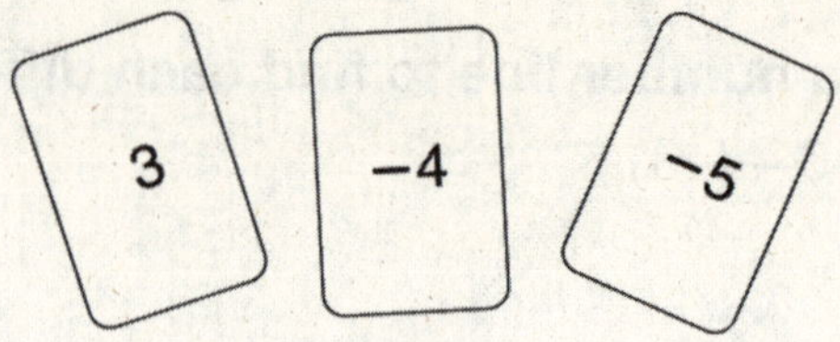

What if you **take away** the 3 card?
Cards –4 and –5 are left.
The new value is –9.
–6 – 3 = –9

What if you **take away** the –4 card?
Cards 3 and –5 are left.
The new value is –2.
–6 – (–4) = –2

Answer each question.

1. Suppose you have the cards shown.
   The total value of the cards is 12.

   a. What if you take away the 7 card? 12 – 7 = ______________
   b. What if you take away the 13 card? 12 – 13 = ______________
   c. What if you take away the –8 card? 12 – (–8) = ______________

2. Subtract –4 – (–2).
   a. –4 < –2. Will the answer be positive or negative? ______________
   b. | 4 | – | 2 | = ______________
   c. –4 – (–2) = ______________

3. Subtract 21 – 13.
   a. 21 > 13. Will the answer be positive or negative? ______________
   b. | 21 | – | 13 | = ______________
   c. 21 – 13 = ______________

Subtract.

4. 31 – (–9) = _____
5. 15 – 18 = _____
6. –9 – 17 = _____
7. –8 – (–8) = _____
8. 29 – (–2) = _____
9. 13 – 18 = _____

Name ____________________ Date ____________ Class ____________

LESSON 2-3

Challenge

Subtract-a-Fact

Complete each temperature and elevation fact. Write the answer in the tables below.

Temperature Facts

1. Louisiana's lowest recorded temperature is (–15 – 1)°F.
2. The highest recorded temperature for Texas is [–18 – (–138)]°F.
3. The normal January temperature for Fairbanks, Alaska, is (–5 – 5)°F.
4. The normal July temperature for Tampa, Florida, is [45 – (–37)]°F.
5. The lowest temperature of 2000 in Detroit, Michigan, was (87 – 90)°F.
6. The highest temperature of 2000 in Norfolk, Virginia, was [–42 – (–138)]°F.
7. North Carolina's lowest recorded temperature is (57 – 91)°F.
8. South Carolina's lowest recorded temperature is (–4 – 15)°F.

Elevation Facts

9. Death Valley's elevation is (–25 – 257) ft.
10. The Caspian Sea's elevation is (125 – 217) ft.
11. Mount Everest's elevation is [–26 – (–29,061)] ft.
12. Mount McKinley's elevation is Mount Everest's elevation – 8,715 ft.
13. Lake Eyre's elevation is (68 – 120) ft.
14. The Dead Sea's elevation is (–761 – 587) ft.
15. Mount Elbrus's elevation is [11,500 – (–7,010)] ft.
16. Lake Assal's elevation is (–254 – 258) ft.

Temperature Facts	°F
Louisiana's lowest	
Texas' highest	
Fairbanks normal January	
Tampa normal July	
Lowest in Detroit in 2000	
Highest in Norfolk in 2000	
North Carolina's lowest	
South Carolina's lowest	

Highest and Lowest Points on Continents	Elevation
Death Valley, North America	
Caspian Sea, Europe	
Mount Everest, Asia	
Mount McKinley, North America	
Lake Eyre, Australia	
Dead Sea, Asia	
Mount Elbrus, Europe	
Lake Assal, Africa	

Name ______________________ Date ______________ Class ____________

LESSON 2-3

Problem Solving

Subtracting Integers

Write the correct answer.

1. The daytime temperature on Mercury can reach 430 °C. The nighttime temperature can drop to –180 °C. How much can the temperature drop during one day?

   ____

2. An ice cream company reported a net profit of $24,000 in 2002 and a net loss of $11,000 in 2003. How much did the company's profits change from 2002 to 2003?

   ____

3. A small business reported a net loss of $86,000 in 2004 and a net profit of $32,000 in 2005. How much did the company's profits change from 2004 to 2005?

   ____

4. The daytime high temperature on the Moon can reach 130 °C. At night, the temperature can drop to –110 °C. What is the difference between the high and low temperatures?

   ____

Choose the letter for the best answer.

5. The low point of the Tonga Trench, in the Pacific Ocean, is –10,630 meters. The low point of the Mariana Trench, also in the Pacific Ocean, is 890 meters lower. What is the depth of the Mariana Trench?

   A 10,630 meters
   B –11,520 meters
   C –9,740 meters
   D 9,740 meters

6. On Wednesday night in St. Petersburg, Russia, the temperature is –11°C. On the same night in Bombay, India, the temperature is 17°C. What is the difference in temperature?

   F –6 °C
   G 50 °C
   H –187 °C
   J 28 °C

7. Climax, Colorado, is the highest town in the United States at 11,560 feet. The lowest town is Calipatria, California, which is 185 feet below sea level. What is the difference in elevation?

   A –185 feet
   B 11,375 feet
   C 11,560 feet
   D 11,745 feet

8. The low point of the Japanese Trench, in the Pacific Ocean, is –10,372 meters. The low point of the Puerto Rico Trench, in the Atlantic Ocean, is 1,172 meters higher. What is the depth of the Puerto Rico Trench?

   F –9,200 meters
   G 8,200 meters
   H –1,172 meters
   J –11,544 meters

Name ______________________ Date ______________ Class ______________

LESSON 2-3

Reading Strategies

Use Graphic Aids

Brett borrowed $7 from his father to buy a CD. He paid back $3. How much money does Brett have now? The number line will help you picture this problem.

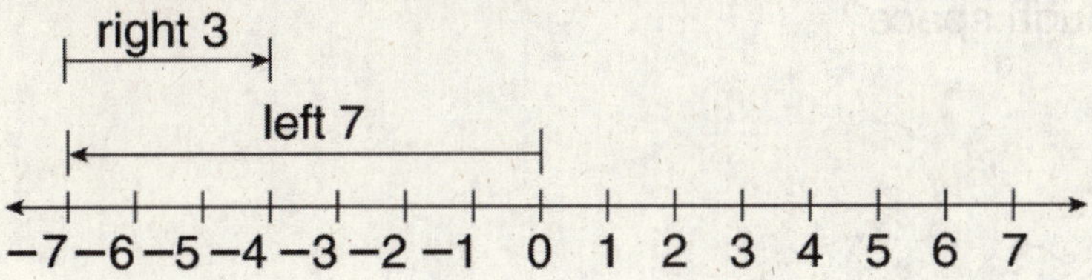

$-7 + 3 = -4$

1. Beginning at 0, in which direction will you move first? ______________

2. How many places? ______________

3. Which direction do you move next? ______________

4. How many places? ______________

5. On what number do you end? ______________

Bret does not have any money. He owes his dad $4. He has negative $4.

Sally and her friends made up a game with points. You can either win or lose up to ten points on each round of the game. After the first round, Sally's team had 2 points. In the second round they lost 6 points. How many points was Sally's team down by after the second round?

The number line will help you picture the problem.

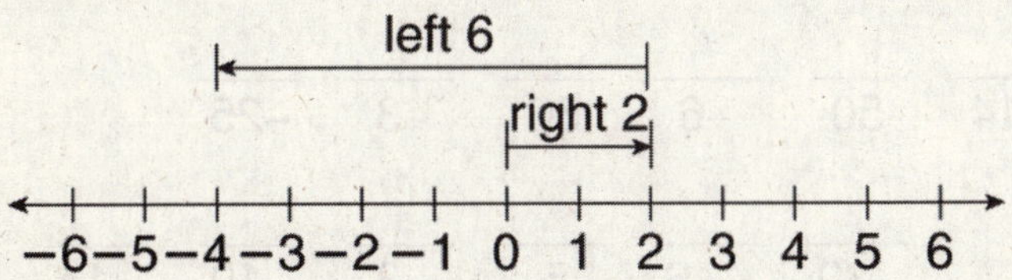

$2 - 6 = -4$

6. Beginning at zero, which direction will you move first? How many places?

____

7. Which direction will you move next? How many places?

____

8. By how many points was Sally's team down? ______________

Name ______________________ Date ____________ Class ____________

LESSON 2-3

Puzzles, Twisters & Teasers

Integer Invasion!

Subtract each pair of integers. Write the letter on the line above the correct answer at the bottom of the page to answer the following question:

What do you call a pan spinning through space?

C −12 − (−2)

A −1 − 8

J −3 − (−8)

N 6 − 10

B −31 − 15

U 12 − (−6)

O −15 − 15

I 30 − (−20)

G −22 − (−5)

D −4 − 21

Y 27 − 19

E −10 − (−7)

R 2 − 9

T −5 − 9

F −8 − (−2)

___ ___
−9 −4

___ ___ ___ ___ ___ ___ ___ ___ ___ ___ ___ ___
18 −4 50 −25 −3 −4 −14 50 −6 50 −3 −25

___ ___ ___ ___ ___ ___   ___ ___ ___ ___ ___ ___
−6 −7 8 50 −4 −17   −30 −46 5 −3 −10 −14

Name ______________________ Date ______________ Class ______________

LESSON 2-4

Practice A

Multiplying and Dividing Integers

Find each product.

1. 6 • (–1) ________
2. –4 • 2 ________
3. –3 • (–4) ________
4. –2 • 8 ________
5. 5 • (–7) ________
6. –7 • 9 ________
7. 8 • 4 ________
8. –3 • (–5) ________
9. –5 • (–5) ________
10. 8 • (–4) ________
11. –7 • (–6) ________
12. 9 • (–8) ________
13. 1 • (–7) ________
14. –4 • (–5) ________
15. –6 • 3 ________
16. –7 • (–7) ________

Find each quotient.

17. 12 ÷ (–4) ________
18. –15 ÷ (–3) ________
19. –20 ÷ 5 ________
20. –27 ÷ (–9) ________
21. –45 ÷ (–5) ________
22. –18 ÷ 9 ________
23. 24 ÷ (–4) ________
24. 32 ÷ 4 ________
25. 21 ÷ 3 ________
26. –36 ÷ (–4) ________
27. 16 ÷ (–4) ________
28. –56 ÷ 8 ________
29. –42 ÷ 7 ________
30. –30 ÷ (–6) ________
31. 27 ÷ 9 ________
32. 25 ÷ 0 ________

33. Jenny was walking down a wooded path. For 4 minutes, the elevation dropped uniformly. Altogether it dropped 12 feet. What was the change in elevation during the first minute?

____

34. As a cold front passed, the temperature dropped steadily 2 degrees each hour. It dropped a total of 10 degrees. How many hours did it take for the temperature to drop this far?

____

Name ______________________ Date ______________ Class ______________

LESSON 2-4

Practice B

Multiplying and Dividing Integers

Find each product.

1. 8 • (–5) ________
2. –4 • 7 ________
3. –6 • (–3) ________
4. –2 • 4 ________
5. 4 • (–9) ________
6. –9 • 5 ________
7. 6 • 8 ________
8. –7 • (–3) ________
9. –6 • (–6) ________
10. 9 • (–3) ________
11. –2 • (–8) ________
12. 5 • (–7) ________
13. 10 • 8 ________
14. –5 • 9 ________
15. 9 • (–6) ________
16. (–4) • (–11) ________

Find each quotient.

17. 25 ÷ (–5) ________
18. –54 ÷ (–6) ________
19. –10 ÷ 5 ________
20. –28 ÷ (–4) ________
21. –42 ÷ (–7) ________
22. –21 ÷ 0 ________
23. 36 ÷ (–6) ________
24. –81 ÷ (–9) ________
25. –32 ÷ 8 ________
26. 45 ÷ (–9) ________
27. –72 ÷ (–8) ________
28. 50 ÷ 10 ________
29. –42 ÷ 6 ________
30. –72 ÷ (–9) ________
31. 40 ÷ 8 ________
32. 56 ÷ 0 ________

33. Kim was walking down a rocky path. For 4 minutes, the elevation dropped uniformly. Altogether it dropped 8 feet. What was the change in elevation per minute for the 4 minutes?

____

34. As a front passed, the temperature changed steadily over 6 hours. Altogether it changed –18 degrees. What was the change in temperature per hour for the 6 hours?

____

Name ____________________ Date ____________ Class ____________

LESSON 2-4

Practice C

Multiplying and Dividing Integers

Find each product.

1. 9 • (–8) ________
2. –7 • 7 ________
3. –5 • (–6) ________
4. –4 • 8 ________
5. 12 • (–10) ________
6. –50 • 3 ________
7. 8 • 10 ________
8. –20 • (–6) ________
9. –7 • (–5) ________
10. 10 • (–9) ________
11. –8 • (–8) ________
12. 12 • (–3) ________
13. –9 • 8 ________
14. 15 • (–4) ________
15. –5 • (–25) ________
16. –150 • 2 ________

Find each quotient.

17. 63 ÷ (–7) ________
18. –48 ÷ (–4) ________
19. –36 ÷ 9 ________
20. –81 ÷ (–9) ________
21. –56 ÷ (–7) ________
22. –54 ÷ 0 ________
23. 80 ÷ (–10) ________
24. –60 ÷ (–12) ________
25. 33 ÷ (–3) ________
26. 100 ÷ (–20) ________
27. –75 ÷ (–3) ________
28. 96 ÷ 8 ________
29. –200 ÷ 50 ________
30. –80 ÷ (–4) ________
31. 48 ÷ 12 ________
32. 77 ÷ 0 ________

33. As Randy swam underwater, his elevation changed uniformly for 4 minutes. Altogether it changed –20 feet. What was the change in elevation during each minute?

____

34. Marie withdrew the same amount of money from her savings account each month for 5 months. Altogether she withdrew $300. She did not make any deposits. What was the change in Marie's savings account balance each month?

____

Name ____________________ Date ______________ Class ______________

LESSON 2-4

Review for Mastery

Multiplying and Dividing Integers

Look for the patterns in these products and quotients.

$1 \cdot 3 = 3$	$-1 \cdot 3 = -3$	$3 \div 1 = 3$	$3 \div (-1) = -3$
$2 \cdot 3 = 6$	$-2 \cdot 3 = -6$	$6 \div 2 = 3$	$6 \div (-2) = -3$
$-3 \cdot (-3) = 9$	$3 \cdot (-3) = -9$	$-9 \div (-3) = 3$	$-9 \div 3 = -3$
$-4 \cdot (-3) = 12$	$4 \cdot (-3) = -12$	$-12 \div (-4) = 3$	$-12 \div 4 = -3$

Look at how to find the signs of the products.

- The product of two integers with the **same sign** is **positive.**

  $(+) \cdot (+) = (+)$ $\qquad$ $(-) \cdot (-) = (+)$

- The product of two integers with **different signs** is **negative.**

  $(+) \cdot (-) = (-)$ $\qquad$ $(-) \cdot (+) = (-)$

Look at how to find the signs of the quotients.

- The quotient of two integers with the **same sign** is **positive.**

  $(+) \div (+) = (+)$ $\qquad$ $(-) \div (-) = (+)$

- The quotient of two integers with **different signs** is **negative.**

  $(+) \div (-) = (-)$ $\qquad$ $(-) \div (+) = (-)$

Find each product or quotient.

1. $-5 \cdot 4$ ________
2. $2 \cdot (-8)$ ________
3. $-1 \cdot (-1)$ ________
4. $-6 \cdot 3$ ________
5. $7 \cdot (-3)$ ________
6. $-8 \cdot (-4)$ ________
7. $-6 \cdot 5$ ________
8. $-9 \cdot (-9)$ ________
9. $36 \div (-4)$ ________
10. $-27 \div 9$ ________
11. $-24 \div (-6)$ ________
12. $-30 \div 5$ ________
13. $18 \div 6$ ________
14. $32 \div (-8)$ ________
15. $-45 \div 9$ ________
16. $-40 \div (-10)$ ________

Name ______________________ Date ______________ Class ______________

LESSON 2-4

Challenge

Football Follies

Phil the Fumbler and Sack'im Sam play for two different football teams. Last year, they both had their worst season ever. Read their interview with a local sports announcer. Find out which player had the worse year by completing the table below.

Announcer: Well Phil, you didn't have a very good season, did you?

Phil: No, I guess losing 42 yards in each of the first 6 games was not what I had hoped to do.

Sam: Gee, that wasn't a very good opening, Phil. At least I gained 53 yards in each of my first 2 games, although, I did lose 88 yards in each of the next 4 games.

Announcer: Yes, Sam, that's true. Now can you tell us how you did in games 7 through 12 last season?

Sam: I can, but I don't really want to.

Announcer: I heard you were sacked and lost 125 yards in each of those games. Is this accurate?

Sam: Yup.

Announcer: Phil, how did you do in games 7 through 12?

Phil: Not that bad. I fumbled quite a bit in games 7 through 11, so I lost 130 yards in each of those games. But I came back in game 12 and gained 85 yards during that one.

Announcer: And what was your yardage for games 13 through 16?

Phil: I lost 95 yards in each game.

Sam: Me, too.

Announcer: Well, I sure hope next season is better for both of you.

Phil and Sam: ME, TOO!

Complete the table.

	Phil	Sam
Total Yards Gained		
Total Yards Lost		
Total Yards		
Average per Game for 16 Games (to the nearest tenth)		
Which player had the worse year?		

Name ______________________ Date ______________ Class ______________

LESSON 2-4

Problem Solving

Multiplying and Dividing Integers

Gerald recorded the temperature at 6-hour intervals for one 24-hour period. The chart shows his data.

Time	10:00 P.M.	4:00 A.M.	10:00 A.M.	4:00 P.M.	10:00 P.M.
Temperature (°F)	70	58	64	82	76

Write the correct answer.

1. If the temperature changed steadily, what was the change per hour from 10:00 P.M. the first day to 4:00 A.M.?

____

2. If the temperature changed steadily, what was the change per hour from 4:00 A.M. to 10:00 A.M.?

____

3. If the temperature changed steadily, what was the change per hour from 10:00 A.M. to 4:00 P.M.?

____

4. If the temperature changed steadily, what was the change per hour from 4:00 P.M. to 10:00 P.M. the second day?

____

Choose the letter for the best answer.

5. A small company had a profit of –$528 in January. If it continues to have the same profit each month for 4 months, what will be the company's total profit for 4 months?

   A –$132 C –$2,112
   B $132 D $2,112

6. Evi's watch is water resistant up to –15 feet. Mateo's watch is water resistant up to 8 times the depth of Evi's watch. Mateo's watch is water resistant to what maximum depth?

   F –7 feet H –120 feet
   G 120 feet J 7 feet

7. A submarine at –235 meters dives to a depth 8 times its initial depth. To what depth does the submarine dive?

   A –227 meters
   B –243 meters
   C –1,880 meters
   D –8,235 meters

8. A 1-kilogram rock dropped into the ocean would take 37 minutes to reach –6,660 meters. If the rock dropped steadily, how far would it fall in 1 minute?

   F 37 meters H 180 meters
   G 167 meters J 188 meters

Name ________________________ Date ______________ Class ______________

LESSON 2-4

Reading Strategies

Use Graphic Aids

The opposite of 5 is negative 5. Owing money is the opposite of having money. Owing $5 is the opposite of having $5.

1. What is the opposite of owing $10? ________________________

2. What is the opposite of having $17? ________________________

David borrowed $4 from his mother each of the last three months. How much money does he owe his mother? The money he owes his mother is a negative number. This problem can be pictured on a number line.

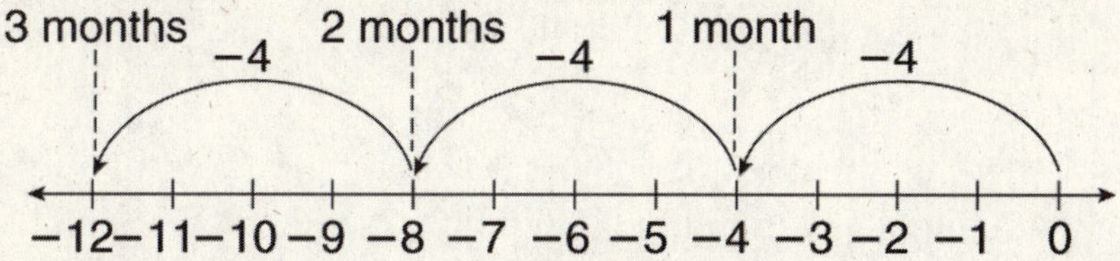

$3 \cdot (-4) = -12$

Use the number line to help you answer the questions.

3. Starting at zero, which direction do you move first? ____________________

4. How many places do you move? ____________________

5. Which direction do you move next? ____________________

6. How many places do you move? ____________________

7. Which direction do you move next? ____________________

8. How many places do you move? ____________________

9. How much money does David owe his mother? ____________________

10. If David borrowed $4 for one more month, how much would he owe his mother? ____________________

Name ______________________ Date ______________ Class ______________

LESSON 2-4

Puzzles, Twisters & Teasers

It's Raining Money!

Solve the equations. Then use the letter associated with the answer to solve the riddle.

H	3 • (–3) =	________
E	4 • 2 =	________
N	4 • (–2) =	________
T	–4 • 3 =	________
I	3 • 3 =	________
S	2 • (–3) =	________
C	–4 • 1 =	________
G	3 • 2 =	________
A	–3 • –4 =	________
R	5 • 2 =	________

When does it rain money?

W ___ ___ ___   ___ ___ ___ ___ ___
–9 8 –8   –12 –9 8 10 8

___ ___   ___ ___ ___ ___ ___ ___   ___ ___
9 –6   –4 –9 12 –8 6 8   9 –8

___ ___ ___   **W** ___ ___ ___ ___ ___ ___
–12 –9 8   8 12 –12 –9 8 10

Name ______________________ Date ______________ Class ______________

LESSON 2-5

Practice A

Solving Equations Containing Integers

Solve each equation. Check your answer.

1. $n - 6 = -2$
2. $x - 8 = -11$
3. $7 = a - 5$
4. $y + 4 = 2$
5. $c + 7 = -3$
6. $0 = v + 1$
7. $8j = -16$
8. $-3k = 24$
9. $-20 = -4s$
10. $\frac{m}{-2} = -5$
11. $\frac{d}{6} = -3$
12. $\frac{r}{-7} = 4$
13. $p + 8 = -6$
14. $-15 = 5b$
15. $f - 9 = -1$
16. $\frac{n}{6} = -4$
17. $k + 10 = 3$
18. $4a = -16$
19. $-6x = -36$
20. $2 = e - 7$
21. $3 = \frac{m}{2}$

22. The temperature in Minnesota was –8 °F one day. This was 12 degrees less than the temperature in Indiana on the same day. What was the temperature in Indiana?

23. Mr. Harding sold 100 shares of stock at $14 per share. He had a loss of $6 per share. What did Mr. Harding pay for each share of the stock?

Name ______________________ Date ____________ Class ____________

LESSON 2-5

Practice B

Solving Equations Containing Integers

Solve each equation. Check your answer.

1. $y - 5 = -4$ ________
2. $n - 9 = -14$ ________
3. $13 = x - 15$ ________
4. $p + 18 = 14$ ________
5. $q + 6 = -2$ ________
6. $0 = w + 4$ ________
7. $9h = -36$ ________
8. $-3b = 36$ ________
9. $-100 = -4u$ ________
10. $\frac{d}{5} = -7$ ________
11. $\frac{c}{4} = -20$ ________
12. $\frac{s}{-9} = 9$ ________
13. $f + 15 = -16$ ________
14. $-75 = 3v$ ________
15. $g - 19 = -21$ ________
16. $-63 = -9s$ ________
17. $14 + m = -10$ ________
18. $12 = \frac{w}{4}$ ________
19. $x = 15 - 31$ ________
20. $\frac{e}{-7} = 8$ ________
21. $-6 = 21 - n$ ________

22. The temperature in Buffalo, New York, was –2 °F one day. This was 42 degrees warmer than the temperature in Nome, Alaska, on the same day. What was the temperature in Nome?

____

23. LaSanda bought 20 shares of stock for $175. She sold the stock for a total profit of $25. What was the selling price of each share of stock?

____

Name ______________________ Date ______________ Class ______________

LESSON 2-5

Practice C

Solving Equations Containing Integers

Solve each equation. Check your answer.

1. $h - 15 = -17$
2. $a - 29 = -12$
3. $13 = g - 20$
4. $m + 48 = 31$
5. $n + 26 = -4$
6. $0 = z + 50$
7. $15y = -75$
8. $0 = -7d$
9. $-144 = -8v$
10. $\frac{k}{-5} = -14$
11. $\frac{r}{4} = -250$
12. $\frac{t}{-11} = 11$
13. $s + 25 = -18$
14. $-96 = -4j$
15. $x - 49 = -57$

Translate each sentence into an equation. Then solve the equation.

16. 8 less than a number n is −13.
17. A number f multiplied by −12 is −72.
18. The sum of −47 and a number d is 84.
19. The product of 8 and a number q is −40.
20. Jake bought 50 shares of stock for $1,000 and sold them for a loss of $375. What was the selling price of each share of stock?
21. Mrs. Lee bought 50 shares of Wiget stock. She paid a total of $1,200, which included a $25 commission fee. Two years later she sold all 50 shares for a total of $1,500. This included a $25 commission fee to the broker. What was the cost and selling price per share of the stock?

Name ______________________ Date ____________ Class ____________

LESSON 2-5

Review for Mastery

Solving Equations Containing Integers

- You can use addition to solve an equation involving subtraction.

  Addition undoes subtraction. Adding the same number to both sides of the equation keeps the equation balanced.

	Check
$x - 5 = -6$	$x - 5 = -6$
$x - 5 + 5 = -6 + 5$	$-1 - 5 \stackrel{?}{=} -6$
$x = -1$	$-6 \stackrel{?}{=} -6$ ✓

- You can use subtraction to solve an equation involving addition.

  Subtraction undoes addition. Subtracting the same number from both sides of the equation keeps the equation balanced.

	Check
$n + 4 = -15$	$n + 4 = -15$
$n + 4 - 4 = -15 - 4$	$-19 + 4 \stackrel{?}{=} -15$
$n = -19$	$-15 \stackrel{?}{=} -15$ ✓

Solve. Check your answer.

1. $p - 9 = -3$

   $p - 9 +$ ____ $= -3 +$ ____

   ____

2. $w - 2 = -14$

   $w - 2 +$ ____ $= -14 +$ ____

   ____

3. $x - 12 = -5$

   $x - 12 +$ ____ $= -5 +$ ____

   ____

4. $f - 8 = 6$

   $f - 8 +$ ____ $= 6 +$ ____

   ____

5. $6 = m - 7$ ____________

6. $-4 = s - 10$ ____________

7. $-8 = y - 2$ ____________

8. $a + 19 = 7$ ____________

9. $b + 15 = -9$ ____________

10. $39 + t = 45$ ____________

11. $-5 = x + 7$ ____________

12. $-2 = k + 11$ ____________

13. $10 = -3 + j$ ____________

Name ______________________ Date ______________ Class ______________

LESSON 2-5

Review for Mastery

Solving Equations Containing Integers (continued)

- You can use division to solve an equation involving multiplication. Division undoes multiplication. Dividing both sides of the equation by the same number keeps the equation balanced.

	Check
$3y = -9$	$3y = -9$
$\frac{3y}{3} = \frac{-9}{3}$	$3 \cdot (-3) \stackrel{?}{=} -9$
$y = -3$	$-9 \stackrel{?}{=} -9$ ✓

- You can use multiplication to solve an equation involving division. Multiplication undoes division. Multiplying both sides of an equation by the same number keeps the equation balanced.

	Check
$\frac{a}{-5} = -8$	$\frac{a}{-5} = -8$
$-5 \cdot \frac{a}{-5} = -8 \cdot (-5)$	$\frac{40}{-5} \stackrel{?}{=} -8$
$a = 40$	$-8 \stackrel{?}{=} -8$ ✓

Solve. Check your answer.

14. $5g = -35$

$\frac{5g}{\_\_\_} = \frac{-35}{\_\_\_}$

____

15. $-8y = -96$

$\frac{-8y}{\_\_\_} = \frac{-96}{\_\_\_}$

____

16. $54 = -6f$

$\frac{54}{\_\_\_} = \frac{-6f}{\_\_\_}$

____

17. $3e = -33$ ______________

18. $-49 = 7n$ ______________

19. $-75 = -5c$ ______________

20. $\frac{n}{4} = -15$ ______________

21. $\frac{m}{-6} = -9$ ______________

22. $\frac{s}{-10} = 8$ ______________

23. $4 = \frac{w}{-6}$ ______________

24. $9 = \frac{z}{5}$ ______________

25. $-11 = \frac{h}{6}$ ______________

Name ______________________ Date ______________ Class ______________

LESSON 2-5

Challenge

Balloon Trail

Match each statement with one of the equations A through J. Write the letter of the equation in the space provided. Then solve each equation and write the solution. To find the way to the party, draw a path through the balloons that show the correct solutions to Exercises 1 to 10 in order.

A. $-7 + x = -3$ B. $\frac{n}{4} = -8$ C. $3y = -36$ D. $x - 9 = -4$

E. $-6n = 42$ F. $\frac{y}{5} = 10$ G. $x + 12 = -5$ H. $n - 6 = 2$

I. $-9y = -27$ J. $\frac{y}{-8} = -2$

1. Three times a number is –36. ______________
2. The sum of a number and 12 is –5. ______________
3. A number decreased by 9 is –4. ______________
4. A number divided by 4 is –8. ______________
5. A number divided by 5 is 10. ______________
6. A number divided by –8 is –2. ______________
7. 6 less than a number is 2. ______________
8. The product of a number and –6 is 42. ______________
9. Add –7 to a number. The sum is –3. ______________
10. –9 times a number is –27. ______________

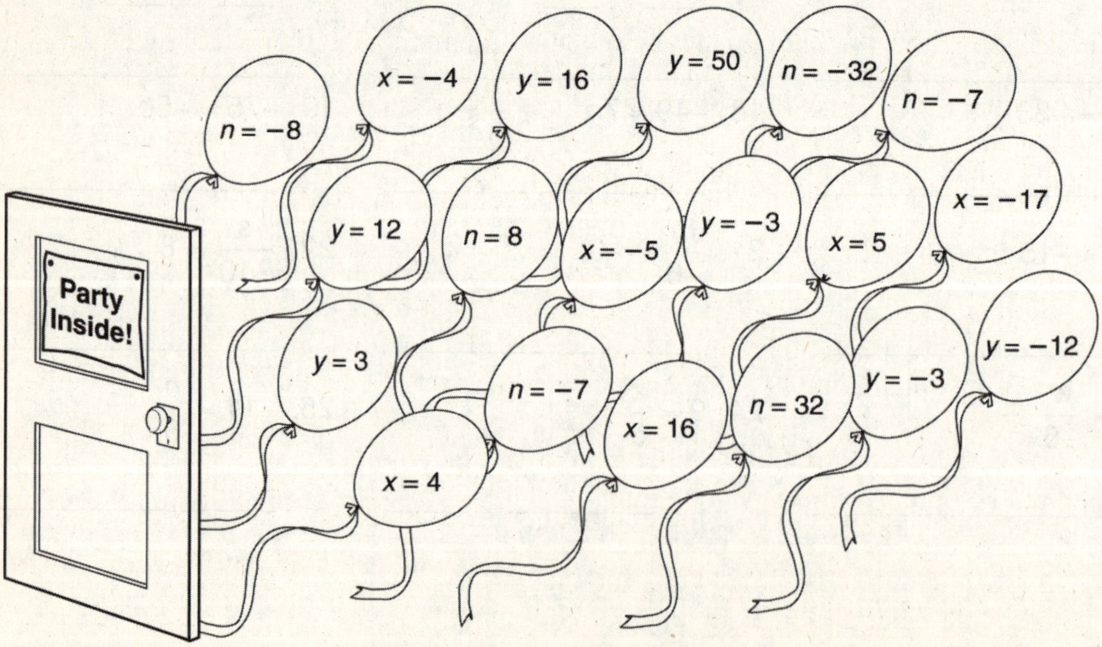

Name ______________________ Date ______________ Class ______________

LESSON 2-5

Problem Solving

Solving Equations Containing Integers

Write the correct answer.

1. Jolene has 30 days to prepare for a bicycle race. She will bicycle 15 miles each day. How many miles will Jolene have bicycled during her training?

____

2. When the amount of money spent is greater than the amount of income, it is called a deficit. In 2002, the U.S. budget deficit was –$158 billion. By 2004, it was –$412 billion. How much did the deficit go up during those two years?

____

3. Two angles of a triangle have a sum of 110°. The sum of all three angles is 180°. What is the measure of the third angle?

____

____

4. The Statue of Liberty was erected in New York City in 1886. How old is the statue?

____

____

Choose the letter for the best answer.

5. Lonnie swam 3 miles each day for 45 days. How many miles did Lonnie swim?
   - A 135 miles
   - B 48 miles
   - C 125 miles
   - D 45 miles

6. On the hottest day in Richmond, Virginia, the temperature was 105 °F. On the coldest day, the temperature was 117 °F lower. What was the coldest temperature in Richmond?
   - F –117 °F
   - G –12 °F
   - H 12 °F
   - J 117 °F

7. Mr. Marco sold 50 shares of Gizmo stock for $1,250. What was the selling price of the stock per share?
   - A $1,200
   - B $250
   - C $50
   - D $25

8. The U.S. Bullion Depository at Fort Knox contains about 147 million troy ounces of gold. At a market price of $400 per troy ounce, what is the approximate value of the gold stored at Fort Knox?
   - F $6 billion
   - G $60 billion
   - H $560 million
   - J $560 billion

Name ______________________ Date ______________ Class ______________

LESSON 2-5

Reading Strategies

Use a Flowchart

The rules for solving equations with integers are the same as with whole numbers.

Use a flowchart to help you follow the rules.

1		2		3		4
Get the variable by itself.	→	Perform the same operation on both sides of the equation.	→	Apply rules for computing integers.	→	Check the solution.

$x + 12 = 5$ → Get x by itself.

$x + 12 - \mathbf{12} = 5 - \mathbf{12}$ → Subtract 12 from both sides.

$x = \mathbf{5} + (-\mathbf{12})$ → Add the opposite.

$x = -7$

$x + 12 = 5$ → Check.

$-7 + 12 \stackrel{?}{=} 5$

$5 = 5$

Use $w - 12 = (-4)$ to answer Exercises 1–4.

1. What operation is used in this equation?

____

2. What operation will you use to get the variable by itself?

____

3. Apply this operation to both sides of the equation.

____

4. What is the value of w?

____

Use $x + (-9) = (-4)$ to answer Exercises 5 and 6.

5. What operation is used in this problem?

____

6. What operation will you use to get x by itself?

____

Name ______________________________ Date ________________ Class ______________

LESSON 2-5

Puzzles, Twisters & Teasers

A Bold Conclusion!

Solve each equation. Then use the boldfaced letters to answer the riddle.

Q $9y = 900$ ____________

A $d - 15 = 45$ ____________

U $j + 56 = -7$ ____________

C $v - 39 = -16$ ____________

F $w + 41 = 0$ ____________

K $-85 = -5c$ ____________

X $11y = -121$ ____________

E $-23 = x + 35$ ____________

O $200 = -25p$ ____________

R $-20 = 2z$ ____________

B $a - 25 = -5$ ____________

S $s + 2 = -4$ ____________

W $-5a = 300$ ____________

D $k = -18 + 11$ ____________

G $b - 7 = -16$ ____________

What do you call ten ducks in a cardboard box?

___   ___ ___ ___   ___ ___

60   20 –8 –11   –8 –41

___ ___ ___ ___ ___ ___ ___ ___

100 –63 60 23 17 –58 –10 –6

CHAPTER 2

Family Letter

2B Factors and Multiples

Dear Family,

The student will be learning divisibility rules. Finding the **prime factorization** of a composite number is one way to find its prime factors. A prime number is a whole number greater than 1 that has exactly two factors, 1 and itself.

The student will find the prime factorization of a number by using a factor tree and a step diagram.

Use a factor tree to find the prime factorization of 84.

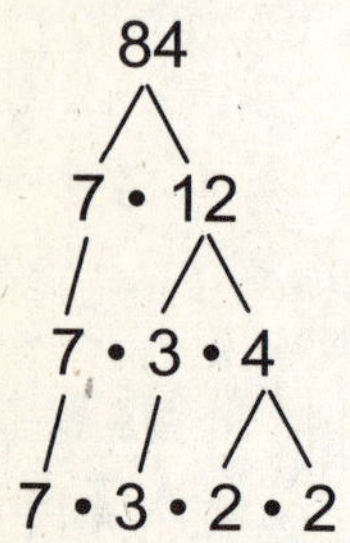

Write 84 as the product of two factors.

Continue factoring until all the numbers are prime

The prime factorization of 84 is 7 • 3 • 2 • 2.
Using exponents, this is written as 22 • 3 • 7.

Use a step diagram to find the prime factorization of 242.

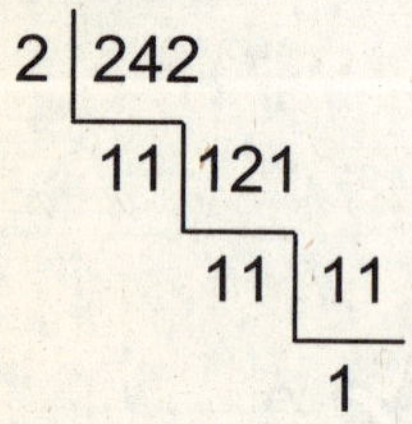

Divide 242 by the smallest prime number possible.
121 is not divisible by 2, so find the next prime number that will divide evenly into 121.
Divide 11 by 11.
The quotient is 1.

The prime factorization of 242 is $2 \cdot 11^2$.

Vocabulary

These are the math words we are learning:

composite number a whole number that has more than two factors

greatest common factor (GCF) the greatest whole number that divides evenly into each number in a group

least common multiple (LCM) the common multiple with the least value

multiple the product of that number and a nonzero whole number

prime factorization when a composite number is written as a product of its prime factors

prime number a whole number that is greater than 1 and that has exactly two factors, 1 and itself

CHAPTER 2

Family Letter

2B Factors and Multiples *continued*

Prime factorization is helpful when writing fractions in simplest form. The student will also use prime factorization to find the **greatest common factor (GCF)** and the **least common multiple (LCM)** of two or more given numbers.

To determine the GCF, the student will list all the factors of each given number and then circle the greatest number that each list has in common. Another method the student will learn is to list the prime factors of each given number, circle the matching factors, then find the product of these common factors.

Find the greatest common factor (GCF) of 24 and 40.

$24 = 2 \cdot 2 \cdot 2 \cdot 3$ $40 = 2 \cdot 2 \cdot 2 \cdot 5$	Write the prime factorization of each number and circle the common factors.
$2 \cdot 2 \cdot 2 = 8$	Multiply the common factors.

The GCF is 8.

A **multiple** of a number is the product of that number and a nonzero whole number. For example, the multiples of 8 are 8, 16, 24, 32, ...

The student will find the least common multiple by listing all the multiples of each number and then finding the least value that is in both lists. A second method of finding the LCM uses prime factorization as shown below.

Find the least common multiple (LCM) of 15 and 35.

$15 = 3 \cdot 5$	Write the prime factorization of each number.
$35 = 7 \cdot 5$	Then circle the common factors.
3, 5, 7	List the factors of both numbers, using the circled factors only once.
$3 \cdot 5 \cdot 7 = 105$	Multiply the factors in the list.

The LCM is 105.

Sincerely,

Name ______________________ Date ______________ Class ______________

CHAPTER 2

At-Home Practice

2B Factors and Multiples

Use a factor tree to find the prime factorization.

1. 88
2. 75
3. 17

Use a step diagram to find the prime factorization.

4. 168
5. 405
6. 280

Find the greatest common factor (GCF).

7. 12, 28 ________
8. 36, 45 ________
9. 36, 48, 108 ________
10. 14, 42, 56 ________
11. 70, 105 ________
12. 38, 57 ________
13. 22, 55, 110 ________
14. 8, 36, 72 ________

Find the least common multiple (LCM).

15. 8, 12 ________
16. 18, 24 ________
17. 6, 15 ________
18. 2, 3, 4 ________
19. 6, 9, 12 ________
20. 4, 10, 20 ________
21. 5, 6, 9 ________
22. 2, 5, 7 ________

Answers: **1.** $2^3 \cdot 11$ **2.** $3 \cdot 5^2$ **3.** 17 **4.** $2^3 \cdot 3 \cdot 7$ **5.** $3^4 \cdot 5$ **6.** $2^3 \cdot 5 \cdot 7$ **7.** 4 **8.** 9 **9.** 12 **10.** 14 **11.** 35 **12.** 19 **13.** 11 **14.** 4 **15.** 24 **16.** 72 **17.** 30 **18.** 12 **19.** 36 **20.** 20 **21.** 90 **22.** 70

Name ______________________________ Date ________________ Class ______________

CHAPTER 2

Family Fun

Multiple Mania

Materials

2 number cubes
Game markers

Directions

- Players take turns rolling a pair of number cubes. Find the *Least Common Multiple* for the numbers that were rolled.
- Place a marker on that number on the game board. If all instances of the least common multiple are already covered, then you lose your turn.
- Use the face value if the cube comes up 2, 3, 4, 5, or 6, but use 9 if a 1 is rolled. If you roll double 6's, you can cover any number on the board.
- The game is over when one player has 4 markers in a row horizontally, vertically, or diagonally.

36	45	8	45	20	35
18	36	20	12	6	10
12	4	25	18	45	15
15	9	10	36	9	4
8	12	20	25	30	15
10	12	30	9	15	8

Name ______________________ Date ____________ Class ____________

LESSON 2-6

Practice A

Prime Factorization

Tell whether each number is prime or composite.

1. 25 ____________
2. 16 ____________
3. 21 ____________
4. 29 ____________

Choose the letter for the best answer.

5. What is the prime factorization of 24?
   A $2^3 \cdot 3$
   B $2^2 \cdot 2$
   C $2^2 \cdot 3^2$
   D $2^3 \cdot 3^2$

6. What is the prime factorization of 50?
   F $3 \cdot 5^2$
   G $3^2 \cdot 5$
   H $2 \cdot 5^2$
   J $3^2 \cdot 5^2$

7. What is the prime factorization of 35?
   A $2 \cdot 3 \cdot 5$
   B $5 \cdot 7$
   C $3^2 \cdot 5$
   D $3^2 \cdot 7$

8. What is the prime factorization of 27?
   F 3^3
   G $2^2 \cdot 3$
   H 9^2
   J $2^2 \cdot 3^2$

Use a factor tree to find the prime factorization.

9. 25 ____________
10. 20 ____________
11. 32 ____________
12. 49 ____________
13. 30 ____________
14. 44 ____________
15. 63 ____________
16. 28 ____________

Use a step diagram to find the prime factorization.

17. 93 ____________
18. 110 ____________
19. 80 ____________
20. 42 ____________
21. 625 ____________
22. 98 ____________
23. 140 ____________
24. 92 ____________

Name ______________________ Date ______________ Class ______________

LESSON 2-6

Practice B

Prime Factorization

Tell whether each number is prime or composite.

1. 33 ________
2. 41 ________
3. 52 ________
4. 79 ________
5. 96 ________
6. 121 ________
7. 83 ________
8. 119 ________

Write the prime factorization of each number.

9. 57 ________
10. 49 ________
11. 88 ________
12. 95 ________
13. 105 ________
14. 98 ________
15. 52 ________
16. 42 ________
17. 68 ________
18. 91 ________
19. 60 ________
20. 72 ________
21. 56 ________
22. 144 ________
23. 370 ________
24. 168 ________
25. 124 ________
26. 515 ________
27. 725 ________
28. 220 ________
29. 450 ________
30. 1,000 ________
31. 1,040 ________
32. 2,500 ________

33. The prime factorization of a number is $3^2 \cdot 5 \cdot 11$. What is the number?

____

Name ________________ Date ____________ Class ____________

LESSON 2-6

Practice C

Prime Factorization

Tell whether each number is prime or composite.

1. 91 ________
2. 117 ________
3. 109 ________
4. 73 ________

Write the prime factorization of each number.

5. 84 ________
6. 343 ________
7. 135 ________
8. 180 ________

Use a step diagram to find the prime factorization.

9. 290 ________
10. 900 ________
11. 575 ________
12. 90 ________
13. 220 ________
14. 350 ________
15. 495 ________
16. 480 ________
17. 2,000 ________
18. 1,040 ________
19. 1,650 ________
20. 3,000 ________

21. The prime factorization of a number is $2^2 \cdot 3^5$. What is the number?

____

22. You can factor 150 as $15 \cdot 10$. List four other ways you can factor 150.

____

23. A number x is a prime factor of 65 and 104. What is the number?

____

Write the composite number for each prime factorization.

24. $2^5 \cdot 5^2$ ________
25. $3^3 \cdot 5 \cdot 11$ ________
26. $4^3 \cdot 7 \cdot 13$ ________
27. $2^8 \cdot 3^5$ ________

28. The prime factors of a number are all the prime numbers between 20 and 30. No factor is repeated. What is the number?

____

Name ______________________ Date ______________ Class ______________

LESSON 2-6

Review for Mastery

Prime Factorization

To write the **prime factorization** of a number, write the number as the product of only prime numbers. A prime number has only two factors, itself and 1.

Use a *factor tree* to find the prime factorization of 18.

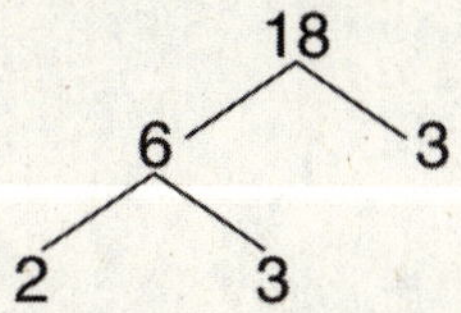

Keep factoring until all the factors are prime.

The prime factorization of 18 is $2 \cdot 3 \cdot 3$, or $2 \cdot 3^2$.

Use a *factor tree* to find the prime factorization of 36.

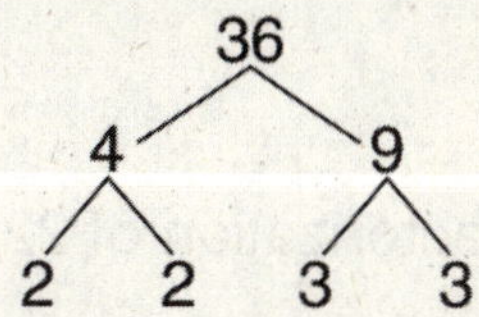

The prime factorization of 36 is $2 \cdot 2 \cdot 3 \cdot 3$, or $2^2 \cdot 3^2$.

Use a factor tree to find the prime factorization.

1. 20 ______________
2. 28 ______________
3. 54 ______________
4. 63 ______________

Use a *step diagram* to find the prime factorization of 60.

2|60
2|30
3|15
5|5
1

1) The divisors must be prime numbers.
2) Keep dividing until the quotient is 1.
3) The divisors are the factors in the prime factorization.

The prime factorization of 60 is $2 \cdot 2 \cdot 3 \cdot 5$, or $2^2 \cdot 3 \cdot 5$.

Use a *step diagram* to find the prime factorization of 75.

3|75
5|25
5|5
1

The prime factorization of 75 is $3 \cdot 5 \cdot 5$, or $3 \cdot 5^2$.

Use a step diagram to find the prime factorization.

5. 48 ______________
6. 24 ______________
7. 40 ______________
8. 98 ______________

Name ______________________ Date ______________ Class ______________

LESSON 2-6

Challenge

A Different Point of View

Create two more factor trees with different factors for each number.

1. 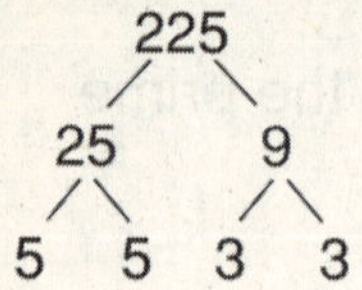

225 225

Prime factorization of 225: ______________

2. 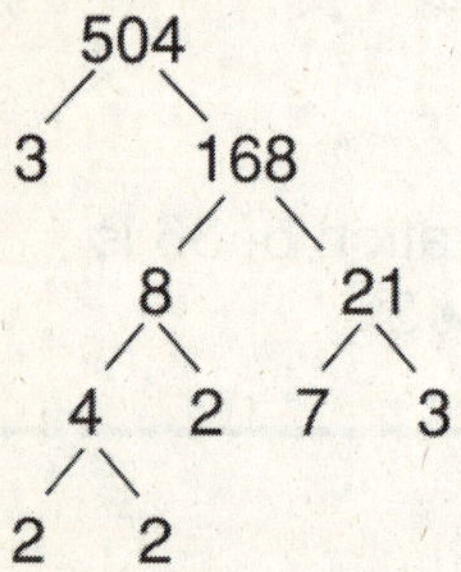

504 504

Prime factorization of 504: ______________

3. In 1742, Christian Goldbach wrote his now famous Goldbach's Conjecture, which states that every even number greater than 2 can be represented as the sum of 2 primes. Although this conjecture is still an open question, it can be supported. Choose 3 even numbers greater than 2 to help support Goldbach's Conjecture.

____

4. Goldbach also made a conjecture that every odd number greater than 7 is the sum of 3 primes. Choose 3 odd numbers to support this conjecture.

____

Name ______________________ Date ______________ Class ______________

LESSON 2-6

Problem Solving

Prime Factorization

Write the correct answer.

1. The width of a swimming pool (in feet) is a prime number greater than 10. The width and length of the pool are factors of 408. What are the dimensions of the pool?

____

2. The area of the dining room at Thomas Jefferson's home in Monticello is about 342 square feet. If the approximate length of one side is a prime number less than 25, what are the approximate dimensions of the room?

____

3. A university has a lounge that can be converted into a meeting hall for 250 people. If the hall is filled and everyone is in equal groups, what are the different ways the people can be grouped so that there are no more than 10 groups?

____

____

4. You want to read a mystery that has 435 pages. If you read the same number of pages per day and the number is a prime number greater than 20, how many pages per day will you read?

____

____

Choose the letter of the best answer.

5. There are 228 seventh graders. Each seventh-grade homeroom starts the year with the same number of students and has at least 15 students. What is the least number of rooms that are needed?

A 17

B 19

C 12

D 21

6. Solve this riddle: I am a number whose prime factors are all the prime numbers between 6 and 15. No factor is repeated. What number am I?

F 9,009

G 91

H 1,001

J 6,006

7. What is the prime factorization of 1,485?

A $3 \cdot 3 \cdot 3 \cdot 5 \cdot 11$

B $3 \cdot 3 \cdot 5 \cdot 5 \cdot 11$

C $3 \cdot 5 \cdot 9 \cdot 11$

D $5 \cdot 11 \cdot 27$

8. Solve this riddle: I am a prime factor of 39 and 65. What number am I?

F 3

G 5

H 11

J 13

Name ______________________ Date ______________ Class ______________

LESSON 2-6

Reading Strategies

Compare and Contrast

Whole numbers are either **prime numbers** or **composite numbers.** Compare the differences between them.

Whole Numbers

Prime Number

- Has only two factors: itself and 1.

Examples

2 → 2 • 1 = 2

5 → 1 • 5 = 5

Composite Number

- Has more than two factors.

Examples

6 → 1 • 6 = 6

2 • 3 = 6

9 → 1 • 9 = 9

3 • 3 = 9

A **composite number** can be written as the product of prime numbers. This is called **prime factorization**.

A **factor tree** helps you find all prime factors of a number.

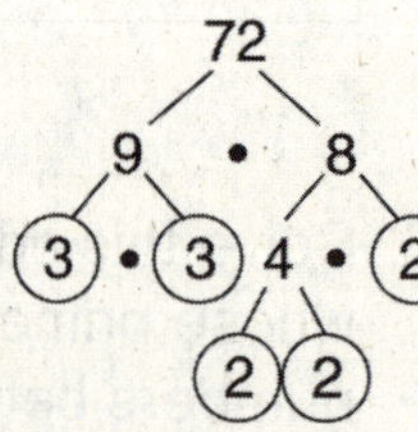

← 9 and 8 are not prime numbers. Continue to factor.

3 is a prime number. → Circle both 3's.

← 2 is a prime number. Circle 2. 4 is not a prime number. Continue to factor.

2 is a prime number. → Circle both 2's.

The prime factorization of 72 is 3 • 3 • 2 • 2 • 2, or $3^2 \cdot 2^3$.

Answer each question.

1. What is the difference between a prime number and a composite number?

____

____

2. Find the prime factorization of 48. ______________________

3. Find the prime factorization of 56. ______________________

Name ______________________ Date ______________ Class ______________

LESSON 2-6

Puzzles, Twisters & Teasers

Can You Find All the Words?

Solve the word search, then find the word from the list that best completes the riddle.

step	unique	composite	factor	factorization
exponent	tree	prime	diagram	Hajratwala

P R I M E F U N I Q U E P
C W E S A H D C B N U O F
T Y P L M A S X D S R W K
E R T Y U J C F T S Y C B
F A C T O R I Z A T I O N
A S Z R A A Q W R E I M F
C G B E X T D S A P B P M
T N M E A W N J I L K O C
O W E R T A T H N B V S I
R Y U I O L C V B N M I L
A S E D I A G R A M F T W
W S C E X P O N E N T E J

What was the mathematician's favorite food?

____ ____ ____ ____ ____ rib

Name ______________________ Date ______________ Class ______________

LESSON 2-7

Practice A

Greatest Common Factor

Draw a line to the greatest common factor (GCF). For Exercises 1–10, choose from A–K; for Exercises 11–20, choose from M–W.

1. 21, 35	A. 3	11. 18, 48	M. 16
2. 12, 32	B. 7	12. 10, 25	N. 60
3. 27, 66	C. 5	13. 120, 180	P. 6
4. 6, 12, 30	D. 4	14. 16, 48, 80	Q. 12
5. 20, 45, 75	E. 9	15. 6, 16, 30	R. 5
6. 27, 36, 45	F. 6	16. 60, 84, 120	S. 3
7. 25, 75, 100	G. 19	17. 42, 63, 84	T. 9
8. 38, 57	H. 8	18. 64, 96	U. 2
9. 32, 40, 120	J. 25	19. 18, 99, 135	V. 21
10. 36, 48, 60	K. 12	20. 24, 45, 90	W. 32

21. A jewelry designer is making identical bracelets. She has 21 blue beads and 14 red beads. What is the greatest number of bracelets she can make using all of the beads?

____

22. A florist is preparing identical flower baskets for a banquet. He has 72 tulips and 60 roses. What is the greatest number of flower baskets he can prepare using all of the tulips and roses?

____

Name _______________ Date _______________ Class _______________

LESSON 2-7

Practice B

Greatest Common Factor

Find the greatest common factor (GCF).

1. 12, 15 ________
2. 22, 33 ________
3. 63, 45 ________
4. 15, 50 ________
5. 18, 81 ________
6. 18, 48 ________
7. 20, 24 ________
8. 14, 42, 49 ________
9. 3, 6, 9 ________
10. 16, 24, 30 ________
11. 16, 40, 88 ________
12. 42, 70 ________
13. 25, 125, 200 ________
14. 26, 39, 52 ________
15. 36, 100 ________
16. 35, 77 ________
17. 56, 84 ________
18. 14, 49, 56, 84 ________
19. 30, 75, 60, 90 ________
20. 12, 38, 40, 94 ________
21. 48, 66, 96, 102 ________
22. Volunteers are preparing identical backpacks for refugees. There are 32 maps and 24 dictionaries to use for the backpacks. What is the greatest number of backpacks they can prepare using all of the maps and dictionaries?

    ____
23. Alyssa is preparing identical fruit baskets. There are 36 oranges and 60 apples to use for the baskets. What is the greatest number of fruit baskets she can prepare using all of the oranges and apples?

    ____

Name ______________________ Date ______________ Class ______________

LESSON 2-7

Practice C

Greatest Common Factor

Find the greatest common factor (GCF).

1. 45, 75 ______
2. 24, 56, 72 ______
3. 14, 18, 22 ______
4. 60, 80, 120 ______
5. 32, 64, 80 ______
6. 32, 40, 120 ______
7. 55, 115 ______
8. 64, 96 ______
9. 14, 35, 56, 77 ______

10. Which pair of numbers has a GCF that is a prime number, 91 and 117 or 84 and 156?

____

11. Which pair of numbers has a GCF that is a prime number, 54 and 99 or 80 and 114?

____

12. Which pair of numbers has a GCF that is a prime number, 70 and 112 or 68 and 187?

____

13. Carvin is preparing identical breakfast gift sets. There are 18 boxes of pancake mix and 27 jars of jam. What is the greatest number of gift sets he can prepare using all of the pancake mix and jam?

____

14. Amelia is decorating identical frames. She has 75 red feathers and 30 blue tiles. What is the greatest number of frames she can decorate using all of the feathers and tiles?

____

15. A jeweler is making identical necklaces. He has 24 gold beads and 144 colored glass beads. What is the greatest number of necklaces he can make using all the gold beads and glass beads?

____

Name ______________________ Date ______________ Class ______________

LESSON 2-7

Review for Mastery

Greatest Common Factor

The **greatest common factor** (GCF) of a group of numbers is the greatest number that can divide each of the numbers without having a remainder.

Use a list to find the GCF of 45, 60, and 75.
List all the factors of each number.

45:	1, 3, 5, 9,		15,		45	
60:	1, 2, 3, 4, 5, 6,	10, 12,	15, 20,	30,	60	
75:	1, 3, 5,		15,	25,		75

The common factors are 1, 3, 5, and 15.
The GCF is 15.

List the factors of each number. Find the greatest common factor (GCF).

1. 9: 1, ____, ____
   24: 1, 2, ____, 4, ____, ____, ____, 24
   GCF = ____

2. 18: 1, 2, 3, ____, ____, ____
   54: 1, 2, 3, ____, ____, ____, ____, 54
   GCF = ____

3. 45: 1, ____, ____, ____, ____, 45
   55: 1, ____, ____, 55
   GCF = ____

4. 12: 1, ____, ____, ____, ____, 12
   20: 1, ____, ____, ____, ____, 20
   GCF = ____

5. 6: 1, ____, ____, ____
   12: 1, 2, ____, ____, ____, 12
   24: 1, 2, ____, ____, ____, ____, 12, 24
   GCF = ____

6. 6: 1, ____, ____, ____
   18: 1, ____, ____, ____, ____, 18
   27: 1, ____, ____, 27
   GCF = ____

Find the greatest common factor (GCF).

7. 27, 45 ____________

8. 14, 18 ____________

9. 12, 30 ____________

10. 10, 45, 70 ____________

Name ______________________ Date ____________ Class ____________

LESSON 2-7

Challenge

The School Trip Factor

There are 24 students in the seventh grade and 36 students in the eighth grade. Solve each problem below about their school trip. Then use your solutions to complete the permission slip each student must have signed in order to go on the trip.

1. The students will visit the zoo, the museum, or the planetarium. Each class will buy its own tickets. Both classes must agree upon the trip that will cost the least. Use the table at right to decide which trip the students will choose.

Trip	Single Tickets	Group Rate	Number of Tickets per Group
Zoo	$8	$24	6
Museum	$5	$12	4
Planetarium	$4	$30	12

2. The principal will provide transportation for both classes. The school PTA will pay the cost but wants the most cost-efficient method of transportation. Use the table at right to decide which method of transportation the PTA will choose.

Type of Transportation	Number of Passengers	Cost
Mini-bus	12	$35
School Bus	70	$185
Mini-van	10	$30

3. The class student council will provide lunch for each class. Each class must agree on the lowest price and then buy at least one lunch pack per student. Use the table at right to decide which lunch deal the students will choose.

Lunch	Number of Lunches per Pack	Cost per Pack
Sandwich and Juice Packs	6	$15
Yogurt and Drink Packs	8	$20
Soup and Sandwich Packs	4	$12

To the Seventh and Eighth-grade Parents:

On Friday, your child will be taking a school trip to the ____________.
The school will pay $___ for the seventh-graders' tickets and $ ___ for the eighth-graders' tickets. The PTA has graciously offered to pay $ ___ for __ [number] ____________ [vehicles] to transport the students. Each class's student council will pay for its class's lunches. Both classes will buy __ [number] ______________ [type] packs at a total cost of $ ___ for the seventh graders and $___ for the eighth graders.

______________________ Parent's signature for students permission

Name ______________________ Date ______________ Class ______________

LESSON 2-7

Problem Solving

Greatest Common Factor

Write the correct answer.

1. Fabric is sold in stores from bolts that are 45 or 60 inches wide. What is the width of the widest strips of fabric you can cut from either bolt without wasting any of the fabric if each strip has the same width?

____

2. The parents are making sandwiches for the class picnic. They have 72 turkey slices, 48 cheese slices, and 96 tomato slices. What is the greatest number of sandwiches they can make if each sandwich has the same filling?

____

3. Two bicycle enthusiasts are leaving Cincinnati at the same time. One is biking 840 miles to Boston. The other is biking 440 miles to Atlanta. What is the greatest number of miles a day each can bike if they want to cover equal distances each day?

____

4. A fruit salad made on a TV cooking program requires chunks of cantaloupe and honeydew. What is the greatest number of servings you can make using all of the fruit if you have 30 chunks of cantaloupe and 42 chunks of honeydew?

____

Choose the letter for the best answer.

5. Ari is making patriotic pins. He has 105 red ribbons, 147 white ribbons, and 189 blue ribbons. What is the greatest number of identical pins he can make if he uses all his ribbons?

   A 52 pins　　C 93 pins
   B 19 pins　　D 21 pins

6. Cheryce is making fruit baskets. She has 60 bananas, 72 pears, 96 apples, and 108 oranges. What is the greatest number of equal baskets she can make with the fruit?

   F 3 baskets　　H 12 baskets
   G 6 baskets　　J 24 baskets

7. There are 100 senators and 435 representatives in the United States Congress. How many identical groups could be formed from all the senators and representatives?

   A 1 group
   B 5 groups
   C 10 groups
   D 15 groups

8. There are 14 baseball teams in the American League and 16 teams in the National League. There are 30 National Hockey League teams. If equal groups of teams are formed, how many hockey teams will be in each group?

   F 3 hockey teams
   G 8 hockey teams
   H 7 hockey teams
   J 15 hockey teams

Name ______________________ Date ______________ Class ______________

LESSON 2-7

Reading Strategies

Follow a Procedure

The **greatest common factor** of two numbers is the largest factor the numbers share or have in common. To find the greatest common factor for 36 and 45, follow these steps.

Step 1: Make a list of the factors for each number. →

36	**45**
1 • 36	1 • 45
2 • 18	3 • 15
3 • 12	5 • 9
4 • 9	
6 • 6	

Step 2: Put the factors in order from smallest to largest.

36: **1,** 2, **3,** 4, 6, **9**, 12, 18, 36
45: **1, 3,** 5, **9,** 15, 45

Step 3: Find the common factors for both numbers. →

1, 3, and **9**

Step 4: Find the greatest common factor for both numbers.

9

Answer each question.

1. What are the factors for 18? ______________
2. Put them in order from least to greatest. ______________
3. What are the factors for 21? ______________
4. Put them in order from least to greatest. ______________
5. Compare your answers for Exercise 2 and Exercise 4. What is the greatest common factor?
6. Make a list of all the factors for 48.

   48

7. Make a list of all the factors for 60.

   60

8. Compare the lists. What factors do both numbers have in common? ______________
9. What is the greatest common factor for 48 and 60? ______________

Name ______________________ Date ______________ Class ______________

LESSON 2-7

Puzzles, Twisters & Teasers

That's the Greatest!

Find the greatest common factor for each set of numbers. Then use the decoder to answer the riddle below.

1. 24, 36, 48, 60 ______________ L
2. 30, 45, 60, 75 ______________ E
3. 20, 40, 50, 120 ______________ K
4. 4, 6, 10, 22 ______________ N
5. 6, 15, 33, 48 ______________ A
6. 14, 21, 35, 70 ______________ R
7. 2, 3, 4, 5, 7 ______________ O
8. 18, 45, 63, 81 ______________ C

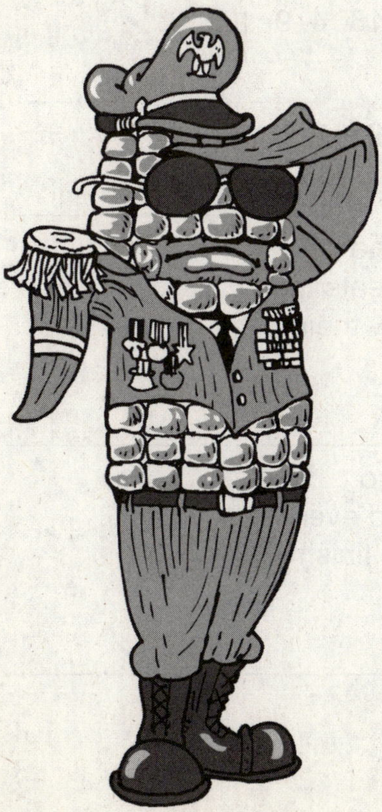

What do you call a piece of corn in the army?

___   ___ ___ ___ ___
3   9   1   7   2

___ ___ ___ ___ ___ ___
10   15   7   2   15   12

Name ______________________ Date ______________ Class ______________

LESSON 2-8

Practice A

Least Common Multiple

Find the correct least common multiple (LCM) for each set of numbers below. Match each LCM with a number in the box.

21	60	33	20	35	75
42	20	40	78	99	36

1. 3, 7 ________
2. 4, 5 ________
3. 10, 12 ________
4. 15, 25 ________
5. 3, 11 ________
6. 9, 12, 18 ________
7. 6, 7, 21 ________
8. 5, 8, 20 ________
9. 2, 26, 39 ________

Find the LCM.

10. 10, 18, 45 ________
11. 5, 6, 20 ________
12. 10, 15, 30 ________
13. 2, 4, 7, 8 ________
14. 3, 9, 12 ________
15. 2, 5, 7, 10 ________
16. 5, 9, 18 ________
17. 4, 10, 12, 15 ________
18. 4, 9, 12, 18 ________
19. 8, 16, 32 ________
20. 5, 15, 20 ________
21. 2, 6, 18, 21 ________

22. Carrie and Lance are playing the drums for the band. In the song, Carrie beats the bass drum every 3 beats. Lance beats the snare drum every 5 beats. On what beat will they first play their drums together?

____

23. Domingo is helping his father paint some boards. Domingo paints a board every 12 minutes. His father paints a board every 8 minutes. If they start together, how long before they will finish painting a board at the same time?

____

Name ______________________ Date ______________ Class ______________

LESSON 2-8

Practice B

Least Common Multiple

Find the least common multiple (LCM).

1. 8, 10 ____________
2. 10, 15 ____________
3. 6, 9 ____________
4. 12, 16 ____________
5. 18, 30 ____________
6. 5, 11 ____________
7. 15, 45 ____________
8. 7, 28 ____________
9. 4, 14 ____________
10. 3, 10, 12 ____________
11. 9, 36, 60 ____________
12. 5, 15 ____________
13. 7, 14, 49 ____________
14. 8, 12, 24, 96 ____________
15. 5, 25, 30 ____________
16. 5, 9, 18 ____________
17. 4, 10, 12, 15 ____________
18. 4, 9, 12, 18 ____________
19. 4, 12, 24, 36 ____________
20. 24, 30, 48, 60 ____________
21. 5, 9, 15, 18 ____________

22. Jasmine is helping her father plant trees to create a border around the back yard. Jasmine plants a tree every 25 minutes, and her father plants a tree every 15 minutes. If they started together, how long before they would finish planting a tree at the same time?

____

23. Two dancers are rehearsing in a studio. One dancer's routine lasts 12 minutes. The other dancer's routine lasts 15 minutes. If they start together and take no breaks between their routines, how long before they start together again?

____

24. Evan and Renzo are swimming laps in the pool. It takes Evan 8 minutes to complete 1 lap and Renzo 6 minutes to complete 1 lap. They start together at the tops of their lanes. In how many minutes will they be together again at the tops of their lanes?

____

Name ______________________ Date ______________ Class ______________

LESSON 2-8

Practice C

Least Common Multiple

Find the least common multiple (LCM).

1. 9, 12 ________
2. 18, 60 ________
3. 9, 15 ________
4. 20, 100 ________
5. 7, 11 ________
6. 8, 12, 16 ________
7. 3, 7, 21 ________
8. 40, 100 ________
9. 5, 9, 18 ________
10. 2, 26, 39 ________
11. 5, 8, 12 ________
12. 20, 40, 50 ________
13. 8, 9, 12 ________
14. 4, 6, 9, 12 ________
15. 3, 9, 12, 72 ________
16. 4, 10, 12 ________
17. 5, 20, 25, 50 ________
18. 4, 7, 8, 28 ________
19. 15, 20, 30 ________
20. 2, 6, 9, 21 ________
21. 10, 32, 64, 80 ________

22. Josh and Eva are packing first-aid supplies. It takes Josh 12 minutes to fill a box with bandages. It takes Eva 8 minutes to fill a box with antibiotics. If they start packing boxes at the same time, how long will it be before they start filling a new box at the same time?

____

23. Sadie is helping her mother stain some shelves. Sadie stains a shelf every 25 minutes. Her mother stains a shelf every 15 minutes. If they start together, how long will it be before they finish staining a shelf at the same time?

____

24. It takes Amanda 14 minutes to climb up the hill and sled down. It takes Alvin 12 minutes to climb up the hill and sled down. If they start up the hill at the same time, how long will it be before they meet again at the starting point?

____

Name ______________________ Date ______________ Class ______________

LESSON 2-8

Review for Mastery

Least Common Multiple

The **least common multiple** (LCM) of two numbers is the least multiple that the two numbers have in common.

- To list the multiples of a number, multiply the number by 1, 2, 3, and so on.

	8 • 1	8 • 2	8 • 3	8 • 4	8 • 5
Multiples of 8	8	16	24	32	40

	6 • 1	6 • 2	6 • 3	6 • 4	6 • 5
Multiples of 6	6	12	18	24	30

- To find the least common multiple of 6 and 8, list several multiples of each number.

Step 1: List multiples of the greater number.

Multiples of 8: 8 16 **24** 32 40

Step 2: List multiples of the lesser number until you find a multiple that is common to each list.

Multiples of 6: 6 12 18 **24**

Since 24 is the first number common to both lists, it is the least common multiple. So the LCM of 6 and 8 is 24.

Find the least common multiple (LCM).

1. 4, 9

   Multiples of 9: 9, 18, 27, 36, 45, 54

   Multiples of 4: 4, 8, 12, 16, 20, 24, 28, 32, 36

   LCM of 4 and 9: _____

2. 6, 7

   Multiples of 7: 7, 14, _____, _____, _____, _____, _____

   Multiples of 6: 6, 12, _____, _____, _____, _____, _____

   LCM of 6 and 7: _____

3. 15, 20

   Multiples of 20: _____, _____, _____, _____, _____

   Multiples of 15: _____, _____, _____, _____

   LCM of 15 and 20: _____

4. 12, 18 _____

5. 16, 20 _____

6. 4, 6, 9 _____

Name ______________________ Date ______________ Class ______________

LESSON 2-8

Challenge

Multiple Solutions

Write the two numbers from each box that have the lowest least common multiple. Then write the LCM.

1.
5	12
6	8

→ Numbers: ____________ LCM: ____________

2.
2	7
9	10

→ Numbers: ____________ LCM: ____________

3.
2	10
5	3

→ Numbers: ____________ LCM: ____________

4.
7	2
10	12

→ Numbers: ____________ LCM: ____________

5.
6	4
7	5

→ Numbers: ____________ LCM: ____________

6.
9	10
15	45

→ Numbers: ____________ LCM: ____________

7.
8	10
6	5

→ Numbers: ____________ LCM: ____________

8.
24	35
32	14

→ Numbers: ____________ LCM: ____________

9.
15	10
7	8

→ Numbers: ____________ LCM: ____________

10.
3	10
12	7

→ Numbers: ____________ LCM: ____________

Use the table at the bottom of the page to solve the riddle. In the middle row, write the LCM of each exercise above the exercise number. Then choose the letter from the Answer Key that matches the LCM and write it in the appropriate box.

Answer Key

C	A	T	E	H
70	6	30	12	10

11. What animal copies from the others on a math test?

9	**4**	**5**		**8**	**2**	**1**	**10**	**6**	**3**	**7**

Name ______________________ Date ______________ Class ______________

LESSON 2-8

Problem Solving

Least Common Multiple

Write the correct answer.

1. Earth revolves around the sun every year. Jupiter revolves around the sun every 12 years. If Earth and Jupiter passed the same point of the sun sometime in 2002, when will they pass that point together again?

____

2. House representatives are elected every 2 years. The President of the United States is elected every 4 years. Both will be elected in 2008. When is the next year after 2008 both will be elected?

____

3. A cat runs a mile every 2 minutes. A squirrel runs a mile every 5 minutes. A cat and a squirrel start together running around a 1-mile track. How long will it be before they meet at the starting point?

____

4. A car manual recommends changing the oil every 5,000 miles and inspecting the engine coolant system every 15,000 miles. At how many miles will both be done together for the first time? for the second time?

____

Choose the letter for the best answer.

5. Mr. Walters receives a dividend every 5 months and a royalty payment every 6 months. He received both in January 2006. When is the next time he would receive both payments in the same month?

   A January 2007
   B April 2008
   C July 2006
   D July 2008

6. Rag Rite Cloth Store always rounds amounts less than whole yards up to the next yard for ribbon purchases. The ribbon that you want to buy comes in rolls of 8 feet. How many rolls should you buy to get the best buy? (Hint: 1 yard = 3 feet)

   F 1 roll
   G 2 rolls
   H 3 rolls
   J 4 rolls

7. The sanitation department picks up recyclable plastics every 3 days. The paper recycling center picks up papers every 4 days. They both picked up on May 4. When will they next pick up on the same day?

   A May 7
   B May 12
   C May 11
   D May 16

8. Hal and Jess both volunteer at the local nursing home. Hal volunteers every 6 days, and Jess volunteers every 8 days. They were both there on Monday. In how many days will they both volunteer together again?

   F in 14 days
   G in 24 days
   H in 48 days
   J in 2 days

Name ______________________________ Date ________________ Class ________________

LESSON 2-8

Reading Strategies

Focus on Vocabulary

A **multiple** of a number is the product of that number and another factor.

Multiples of 5 are the numbers you get when you multiply by **5**.

1 • 5 = **5**   2 • 5 = **10**   3 • 5 = **15**   4 • 5 = **20**

The multiples of 5 listed above are 5, 10, 15, 20.

You can continue to count by 5's to list more multiples of 5: 25, 30, 35, and so on.

Common means something that is shared. A common multiple of two numbers is a multiple that the two numbers share, or have in common.

Multiples of 5 → 5, 10, 15, **20,** 25, 30, 35, **40,** 45, 50, 55, **60**

Multiples of 4 → 4, 8, 12, 16, **20,** 24, 28, 32, 36, **40,** 44, 48, 52, 56, **60**

In the example above, you can see the multiples that are common to both 4 and 5. They are 20, 40, and 60.

The **least common multiple (LCM)** of two numbers is the smallest of the common multiples. 20 is the smallest, or least common, multiple of 4 and 5.

Answer each question.

1. What is a multiple?

____

____

2. What does the word *common* mean?

____

3. What is a common multiple?

____

4. What is the least common multiple?

____

5. Write the first 9 multiples of 2. ______________________

6. Write the first 9 multiples of 3. ______________________

7. Write the common multiples for both 2 and 3. ______________________

8. What is the least common multiple for 2 and 3? ______________________

Name ______________________ Date ______________ Class ______________

LESSON 2-8

Puzzles, Twisters & Teasers

A Common Cause!

Find the least common multiple for each set of numbers. Spell out the answers to solve the crossword puzzle.

Across

1. 3, 5
4. 9, 12
6. 10, 4
7. 2, 5
9. 6, 14

Down

1. 3, 9, 15
2. 4, 6, 24
3. 6, 9
5. 4, 7
6. 3, 7
8. 15, 20

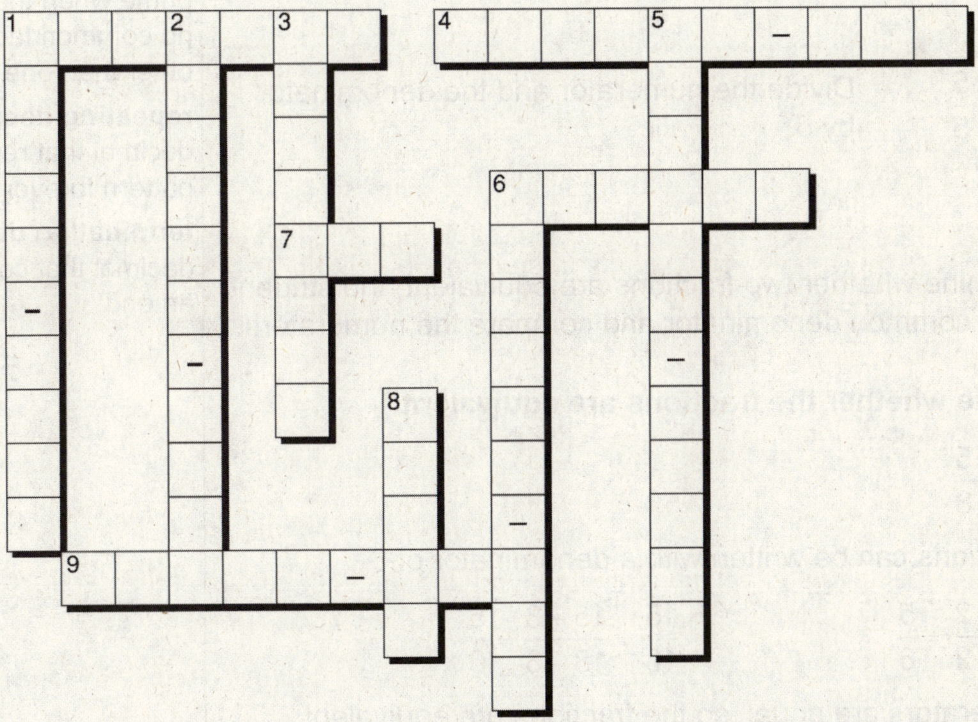

CHAPTER 2

Family Letter

2C Rational Numbers

Dear Family,

The student will be learning about rational numbers. The set of rational numbers includes whole numbers, integers, fractions, and some decimals. Since fractions are such an integral part of rational numbers, the student needs to know how to find equivalent fractions. He or she can create equivalent fractions by multiplying or dividing **both** the numerator and the denominator of a given fraction by the same number.

Find two fractions equivalent to the fraction $\frac{10}{15}$.

$\frac{10}{15} = \frac{10 \cdot 2}{15 \cdot 2}$ Multiply the numerator and the denominator by 2.

$= \frac{20}{30}$

$\frac{10}{15} = \frac{10 \div 5}{15 \div 5}$ Divide the numerator and the denominator by 5.

$= \frac{2}{3}$

To determine whether two fractions are equivalent, the student will find a common denominator and compare the numerators.

Determine whether the fractions are equivalent.

$\frac{10}{12}$ and $\frac{15}{18}$

Both fractions can be written with a denominator of 6.

$\frac{10}{12} = \frac{10 \div 2}{12 \div 2} = \frac{5}{6}$ $\frac{15}{18} = \frac{15 \div 3}{18 \div 3} = \frac{5}{6}$

The numerators are equal, so the fractions are equivalent.

Vocabulary

These are the math words we are learning:

equivalent fractions fractions that name the same amount or part

rational number a number that can be written as a fraction, with integers for its numerator and denominator

relatively prime two numbers are relatively prime when they have no common factors other than one

repeating decimal a decimal that repeats a pattern forever

terminating decimal a decimal that comes to an end

CHAPTER 2

Family Letter

2C Rational Numbers continued

The student will also learn to convert between mixed numbers and improper fractions. This process is helpful when multiplying and dividing fractions, and when simplifying more complex expressions.

Converting from an improper fraction to a mixed number.	**Converting from a mixed number to an improper fraction.**
Write $\frac{17}{5}$ as a mixed number. Divide the numerator by the denominator. "Well, which group is going to $5\overline{)17}$ = 3*R*2 15 2 $\frac{17}{5} = 3\frac{2}{5}$ *Use the quotient and the remainder to write the mixed number.*	**Write $7\frac{4}{9}$ as an improper fraction.** Multiply the denominator and the whole number, and then add the numerator. *Think:* $9 \times 7 + 4 = 67$ $\frac{67}{9}$ *Use the result to write the improper fraction.*

The student will also learn to convert between decimals and fractions. To convert a fraction to a decimal, the student will divide the numerator by the denominator. In some cases, the student might be asked to round this answer to a specified place value.

To convert a decimal to a fraction, the student will need to about place value. The value of the denominator is determined by where the decimal ends. For example in the decimal, 0.45, the "5" is in the hundredths place, so the fractional form of this number is $\frac{45}{100}$.

The student will then reduce this fraction to simplest form by dividing both the numerator and denominator by the greatest common factor, in this case 5. So, $0.45 = \frac{45}{100} = \frac{9}{20}$.

Practice converting between decimals and fractions and converting between mixed numbers and improper fractions with the student.

Sincerely,

Name ______________________ Date ______________ Class ______________

CHAPTER 2

At-Home Practice

2C Rational Numbers

1. Place each number on the number line.

$-\frac{1}{4}$ $5\frac{3}{4}$ $2\frac{1}{2}$ 1.5 $\frac{1}{10}$ −0.75 2.25 $-4\frac{3}{4}$

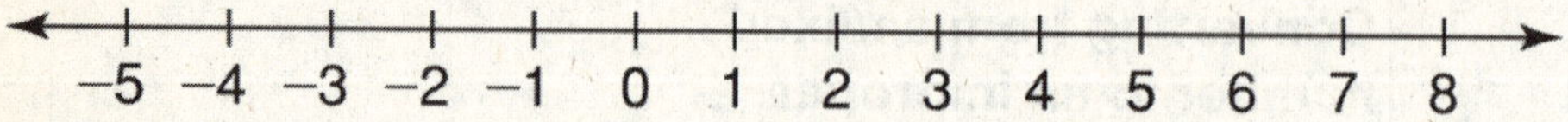

Find two fractions equivalent to each fraction.

2. $\frac{3}{8}$ ______ 3. $\frac{5}{9}$ ______ 4. $\frac{2}{3}$ ______ 5. $\frac{15}{20}$ ______

Convert between improper fractions and mixed numbers.

6. $\frac{17}{6}$ ______ 7. $4\frac{6}{7}$ ______ 8. $\frac{84}{7}$ ______ 9. $3\frac{9}{13}$ ______

Convert each decimal to a fraction in simplest form.

10. 0.85 ______ 11. 3.2 ______ 12. 1.38 ______ 13. 3.07 ______

Compare the fractions using < or >.

14. $\frac{1}{8}$ ___ $\frac{2}{5}$ 15. $\frac{3}{4}$ ___ $\frac{4}{5}$ 16. $\frac{1}{6}$ ___ $\frac{1}{8}$ 17. $\frac{9}{10}$ ___ $\frac{7}{8}$

Place the following numbers in order from least to greatest.

18. 0.6, 0.603, 0.61 ______ 19. −3, 3.01, −2.5 ______ 20. 0.28, 2.08, −2.88 ______

Answers: 1. Each point lies between two whole numbers. **2.** Possible answer: $\frac{6}{16}$, $\frac{9}{24}$ **3.** Possible answer: $\frac{10}{18}$, $\frac{15}{27}$ **4.** Possible answer: $\frac{4}{6}$, $\frac{6}{9}$ **5.** Possible answer: $\frac{3}{4}$, $\frac{30}{40}$ **6.** $2\frac{5}{6}$ **7.** $\frac{34}{7}$ **8.** 12 **9.** $\frac{48}{13}$ **10.** $\frac{17}{20}$ **11.** $3\frac{1}{5}$ **12.** $1\frac{19}{50}$ **13.** $3\frac{7}{100}$ **14.** < **15.** < **16.** > **17.** > **18.** 0.6, 0.603, 0.61 **19.** −3, −2.5, 3.01 **20.** −2.88, 0.28, 2.08

Name ______________________________ Date ______________ Class ______________

CHAPTER 2

Family Fun

Circle Time

Directions

Find pairs of equivalent fractions and/or decimals in the hidden number puzzle. Circle the pairs which appear in a vertical or horizontal line. For each of the 10 pairs below, create a third equivalent fraction or decimal.

$\frac{1}{3}$	$\frac{2}{3}$	$\frac{4}{8}$	$2\frac{1}{2}$	$\frac{5}{2}$	$\frac{4}{5}$	$1\frac{1}{2}$
$\frac{9}{27}$	0.9	0.05	2.2	$\frac{4}{8}$	$1\frac{1}{4}$	1.25
$\frac{15}{20}$	$\frac{7}{8}$	$1\frac{1}{4}$	$\frac{5}{2}$	$\frac{5}{8}$	0.75	$\frac{2}{3}$
0.9	$\frac{9}{10}$	9	$\frac{3}{5}$	$\frac{25}{40}$	0.6	$\frac{36}{48}$
1.5	$\frac{4}{8}$	2	$\frac{22}{2}$	8	0.68	$\frac{6}{8}$
$\frac{18}{3}$	$\frac{7}{5}$	$\frac{16}{8}$	$\frac{4}{9}$	$\frac{11}{3}$	$3\frac{2}{3}$	3.2
9	$\frac{1}{6}$	0.2	$\frac{8}{18}$	$\frac{5}{8}$	0.75	$\frac{15}{20}$

List the 10 equivalent pairs. Create an additional equivalent fraction.

1. ______________________________
2. ______________________________
3. ______________________________
4. ______________________________
5. ______________________________
6. ______________________________
7. ______________________________
8. ______________________________
9. ______________________________
10. ______________________________

Answers: For 1–10 possible equivalent fractions follow answer.

1. $\left(\frac{1}{3};\frac{9}{27}\right);\frac{2}{6}$ **2.** $\left(2\frac{1}{2};\frac{5}{2}\right);\frac{10}{4}$ **3.** $\left(1\frac{1}{4};1.25\right);\frac{10}{8}$ **4.** $\left(0.9;\frac{9}{10}\right);\frac{18}{20}$ **5.** $\left(\frac{36}{48};\frac{6}{8}\right);\frac{12}{16}$ **6.** $\left(2;\frac{16}{8}\right);\frac{8}{4}$

7. $\left(\frac{11}{3};3\frac{2}{3}\right);\frac{22}{6}$ **8.** $\left(0.75;\frac{15}{20}\right);\frac{3}{4}$ **9.** $\left(\frac{4}{9};\frac{8}{18}\right);\frac{12}{27}$ **10.** $\left(\frac{5}{8};\frac{25}{40}\right);\frac{10}{16}$

Name ______________________ Date ______________ Class ______________

LESSON 2-9

Practice A

Equivalent Fractions and Mixed Numbers

Find a fraction equivalent to the given fraction.

1. $\frac{1}{4}$ __________ 2. $\frac{3}{5}$ __________ 3. $\frac{4}{9}$ __________ 4. $\frac{1}{8}$ __________

5. $\frac{1}{6}$ __________ 6. $\frac{20}{40}$ __________ 7. $\frac{6}{15}$ __________ 8. $\frac{35}{50}$ __________

Write the fractions with a common denominator. Then determine if they are equivalent.

9. $\frac{2}{3}$ and $\frac{6}{9}$ __________

10. $\frac{7}{10}$ and $\frac{8}{15}$ __________

11. $\frac{3}{6}$ and $\frac{4}{8}$ __________

12. $\frac{16}{8}$ and $\frac{10}{5}$ __________

13. $\frac{9}{6}$ and $\frac{6}{4}$ __________

14. $\frac{8}{12}$ and $\frac{15}{24}$ __________

Write each as a mixed number.

15. $\frac{5}{4}$ __________ 16. $\frac{9}{2}$ __________ 17. $\frac{13}{5}$ __________ 18. $\frac{12}{7}$ __________

Write each as an improper fraction.

19. $3\frac{2}{3}$ __________ 20. $2\frac{3}{8}$ __________ 21. $1\frac{3}{5}$ __________ 22. $2\frac{5}{6}$ __________

23. Lou's bedroom is $11\frac{1}{2}$ feet long. Write this number as an improper fraction.

____

24. Kaleigh used $\frac{2}{3}$ cup of flour to make cookies. Eric used $\frac{9}{12}$ cup of flour. Did they use the same amount of flour?

____

Name ______________________ Date ______________ Class ______________

LESSON 2-9

Practice B

Equivalent Fractions and Mixed Numbers

Find a fraction equivalent to the given fraction.

1. $\frac{2}{9}$ ________ 2. $\frac{8}{15}$ ________ 3. $\frac{7}{8}$ ________

4. $\frac{16}{24}$ ________ 5. $\frac{12}{20}$ ________ 6. $\frac{9}{12}$ ________

Determine whether the fractions in each pair are equivalent.

7. $\frac{8}{10}$ and $\frac{12}{15}$ ________ 8. $\frac{6}{8}$ and $\frac{8}{12}$ ________ 9. $\frac{3}{9}$ and $\frac{4}{8}$ ________

10. $\frac{7}{4}$ and $\frac{9}{5}$ ________ 11. $\frac{15}{12}$ and $\frac{20}{16}$ ________ 12. $\frac{15}{9}$ and $\frac{30}{18}$ ________

Write each as a mixed number.

13. $\frac{21}{8}$ ________ 14. $\frac{37}{4}$ ________ 15. $\frac{16}{5}$ ________ 16. $\frac{49}{9}$ ________

Write each as an improper fraction.

17. $8\frac{2}{3}$ ________ 18. $1\frac{7}{12}$ ________ 19. $25\frac{3}{4}$ ________ 20. $7\frac{5}{6}$ ________

21. Maria's desk is $33\frac{3}{4}$ inches long. Write this number as an improper fraction.

____

22. Leon walked $\frac{5}{8}$ mile. Liz walked $\frac{10}{16}$ mile. Did they walk the same distance?

____

Name ______________________ Date ______________ Class ______________

LESSON 2-9

Practice C

Equivalent Fractions and Mixed Numbers

Find two fractions equivalent to each number.

1. $\frac{7}{12}$ ________
2. $\frac{11}{15}$ ________
3. $\frac{14}{25}$ ________
4. $\frac{18}{20}$ ________
5. 5 ________
6. $1\frac{1}{8}$ ________

Determine whether the fractions in each pair are equivalent.

7. $\frac{10}{15}$ and $\frac{12}{20}$ ________
8. $\frac{10}{18}$ and $\frac{9}{12}$ ________
9. $\frac{18}{24}$ and $\frac{12}{16}$ ________
10. $\frac{7}{8}$ and $\frac{9}{10}$ ________
11. $\frac{32}{6}$ and $\frac{48}{9}$ ________
12. $\frac{38}{15}$ and $\frac{25}{10}$ ________

Write each as a mixed number.

13. $\frac{43}{7}$ ________
14. $\frac{27}{5}$ ________
15. $\frac{55}{8}$ ________
16. $\frac{80}{11}$ ________

Write each as an improper fraction.

17. $25\frac{2}{5}$ ________
18. $11\frac{7}{9}$ ________
19. $4\frac{11}{12}$ ________
20. $5\frac{5}{16}$ ________

21. Write an improper fraction and a mixed number to represent 139 inches in feet.

____

22. A baker used $52\frac{1}{2}$ cups of flour to make 15 loaves of bread. How many quarter cups of flour were used?

____

Name ______________________ Date ______________ Class ______________

LESSON 2-9

Review for Mastery

Equivalent Fractions and Mixed Numbers

Equivalent fractions name the same number.

- You can use fraction strips to find equivalent fractions.

Find two fractions equivalent to $\frac{6}{8}$.

First, use fraction strips to model $\frac{6}{8}$.

Then find equivalent fractions by placing other fraction strips with the same length below the model.

$\frac{1}{8}$	$\frac{1}{8}$	$\frac{1}{8}$	$\frac{1}{8}$	$\frac{1}{8}$	$\frac{1}{8}$	
$\frac{1}{8}$	$\frac{1}{8}$	$\frac{1}{8}$	$\frac{1}{8}$	$\frac{1}{8}$	$\frac{1}{8}$	$\frac{6}{8}$

$\frac{1}{4}$	$\frac{1}{4}$	$\frac{1}{4}$	$\frac{3}{4}$

$\frac{1}{16}$	$\frac{1}{16}$	$\frac{1}{16}$	$\frac{1}{16}$	$\frac{1}{16}$	$\frac{1}{16}$	$\frac{1}{16}$	$\frac{1}{16}$	$\frac{1}{16}$	$\frac{1}{16}$	$\frac{1}{16}$	$\frac{1}{16}$	$\frac{12}{16}$

$\frac{3}{4}$ and $\frac{12}{16}$ are equivalent to $\frac{6}{8}$.

Use fractions strips to find a fraction equivalent to each fraction.

1. $\frac{1}{2}$ ________ 2. $\frac{1}{4}$ ________ 3. $\frac{2}{5}$ ________ 4. $\frac{2}{3}$ ________

- You can use cross products to determine if two fractions are equivalent.

$\frac{8}{10} \times \frac{12}{15}$ $8 \cdot 15 \stackrel{?}{=} 10 \cdot 12$

$120 = 120$

$\frac{8}{10}$ and $\frac{12}{15}$ are equivalent. Their cross products are equal.

$\frac{9}{15} \times \frac{2}{3}$ $9 \cdot 3 \stackrel{?}{=} 15 \cdot 2$

$27 \neq 30$

$\frac{9}{15}$ and $\frac{2}{3}$ are not equivalent. Their cross products are not equal.

Find the cross products. Determine if the two fractions are equivalent. Write *yes* or *no*.

5. $\frac{18}{6}$ and $\frac{10}{3}$ 6. $\frac{6}{10}$ and $\frac{9}{15}$ 7. $\frac{2}{3}$ and $\frac{9}{12}$

________ ________ ________

Name ______________________ Date ______________ Class ______________

LESSON 2-9

Review for Mastery

Equivalent Fractions and Mixed Numbers (continued)

In an *improper fraction* the numerator is greater than the denominator, such as $\frac{11}{3}$.

A *mixed number* is made up of both a whole number part and a fraction part, such as $2\frac{5}{8}$.

- To write an improper fraction as a mixed number, you can write the improper fraction as a sum of fractions.

$$\frac{11}{3} = \underbrace{\frac{3}{3} + \frac{3}{3} + \frac{3}{3}}_{1\,+\,1\,+\,1} + \frac{2}{3} = \frac{2}{3}$$

- To write a mixed number as an improper fraction, you can write the whole number as a fraction with the same denominator as the fraction part.

$$2\frac{5}{8} = 2 + \frac{5}{8} = \frac{16}{8} + \frac{5}{8} = \frac{21}{8}$$

8. Write $\frac{13}{4}$ as a mixed number.

$$\frac{13}{4} = \underset{\substack{\uparrow \\ 1}}{\frac{__}{4}} + \underset{\substack{\uparrow \\ 1}}{\frac{__}{4}} + \underset{\substack{\uparrow \\ 1}}{\frac{__}{4}} + \frac{__}{4} = ________$$

9. Write $3\frac{2}{5}$ as an improper fraction.

$$3\frac{2}{5} = ________ + \frac{__}{} = \frac{__}{5} + \frac{__}{} = \frac{__}{}$$

Write each improper fraction as a mixed number.

10. $\frac{23}{5}$ ________ 11. $\frac{15}{8}$ ________ 12. $\frac{54}{7}$ ________ 13. $\frac{14}{3}$ ________

Write each mixed number as an improper fraction.

14. $5\frac{1}{2}$ ________ 15. $3\frac{3}{8}$ ________ 16. $2\frac{11}{15}$ ________ 17. $8\frac{3}{4}$ ________

Name ________________________ Date ______________ Class ______________

LESSON 2-9

Challenge

Target Practice

Find out where the dart lands on the dartboard. Cross out the fraction, improper fraction, or mixed number on the dartboard that is equivalent to each number below. The dart lands on the number that is left.

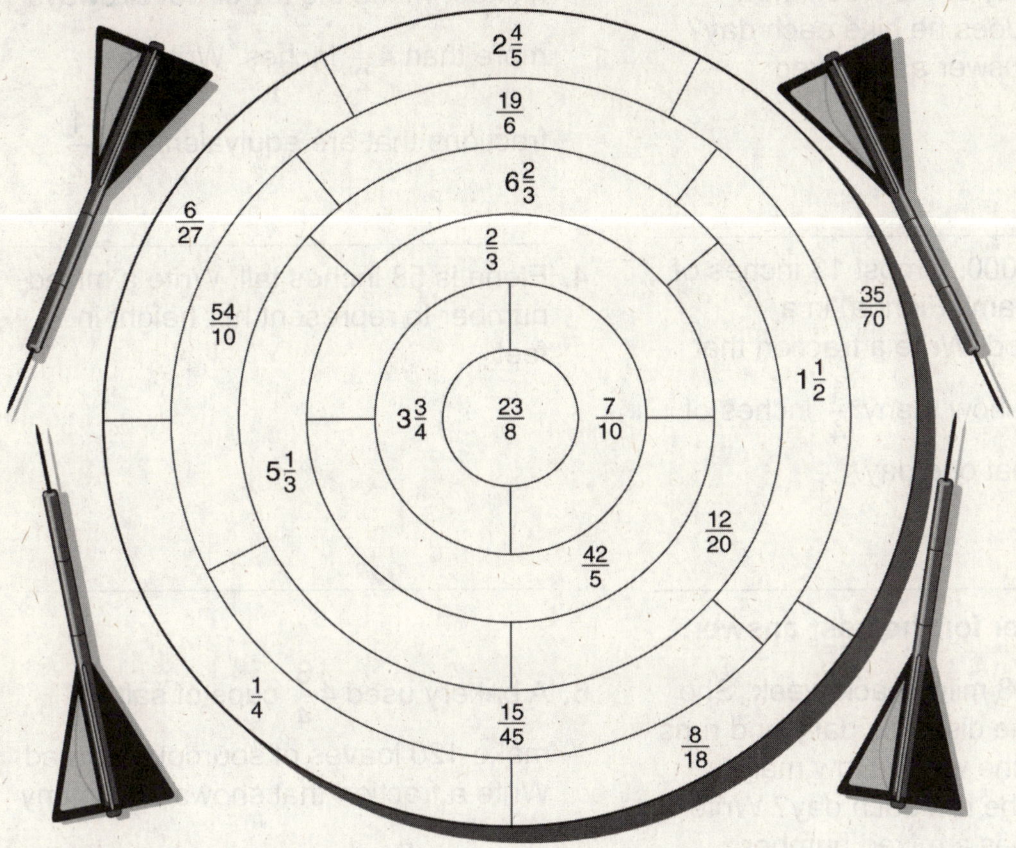

1. $\frac{32}{6}$
2. $\frac{14}{20}$
3. $\frac{12}{8}$
4. $\frac{3}{5}$
5. $2\frac{7}{8}$
6. $\frac{28}{10}$
7. $\frac{16}{36}$
8. $\frac{50}{75}$
9. $\frac{2}{9}$
10. $\frac{20}{3}$
11. $\frac{5}{15}$
12. $3\frac{1}{6}$
13. $\frac{3}{12}$
14. $5\frac{2}{5}$
15. $\frac{45}{12}$
16. $\frac{1}{2}$

17. The dart lands on ______________

Name ______________________ Date ______________ Class ______________

LESSON 2-9

Problem Solving

Equivalent Fractions and Mixed Numbers

Write the correct answer.

1. Bert bicycles 136 miles each week. If he bikes the same distance daily and bikes each day of the week, how many miles does he bike each day? Write your answer as a mixed number.

____

2. The Eiffel Tower in Paris, France, was designed so well that even in the highest winds the tower never sways more than $4\frac{1}{2}$ inches. Write two fractions that are equivalent to $4\frac{1}{2}$.

____

3. In October 2000, almost 13 inches of rain fell in Miami, Florida, in a 24-hour period. Write a fraction that shows about how many $\frac{1}{4}$ inches of rain fell on that one day.

____

4. Elena is 58 inches tall. Write a mixed number to represent her height in feet.

____

Choose the letter for the best answer.

5. Randi runs 48 miles each week. She runs the same distance daily and runs each day of the week. How many miles does she run each day? Write your answer as a mixed number.

A $6\frac{6}{7}$ miles

B $6\frac{5}{6}$ miles

C $6\frac{7}{48}$ miles

D $7\frac{1}{7}$ miles

6. A bakery used $4\frac{3}{4}$ cups of salt to make 120 loaves of sourdough bread. Write a fraction that shows how many $\frac{1}{4}$ cups of salt were used to make the bread.

F $\frac{30}{4}$ cups

G $\frac{24}{4}$ cups

H $\frac{12}{4}$ cups

J $\frac{19}{4}$ cups

7. One of the driest summers in the Northeast was in 1909, with $\frac{896}{100}$ inches of precipitation. Which mixed number is equivalent to $\frac{896}{100}$?

A $89\frac{6}{10}$

B $8\frac{96}{10}$

C $8\frac{24}{25}$

D $8\frac{4}{5}$

8. The Central Ohio Transit Authority has $957\frac{3}{5}$ miles of routes. Which fraction is equivalent to $957\frac{3}{5}$?

F $\frac{4,781}{5}$

G $\frac{4,788}{5}$

H $\frac{4,783}{5}$

J $\frac{9,573}{5}$

Name ______________________ Date ______________ Class ______________

LESSON 2-9

Reading Strategies

Multiple Meaning Terms

If you multiply or divide the numerator and denominator of a fraction by the same number, you get another fraction that is **equivalent,** or equal.

$$\frac{2}{3} \cdot \frac{2}{2} = \frac{4}{6}$$

In the example above, both the numerator and denominator are multiplied by 2. Equal fractions are called **equivalent fractions.**

$\frac{2}{3}$ is equivalent to $\frac{4}{6}$.

Answer the following questions.

1. Is $\frac{1}{2}$ equivalent to $\frac{6}{12}$? ______________

2. How do you know?

____

____

3. Are $\frac{1}{5}$ and $\frac{2}{3}$ equivalent fractions? ______________

4. How do you know?

____

____

5. Is $\frac{7}{15}$ equivalent to $\frac{1}{7}$? ______________

6. How do you know?

____

____

7. Is $\frac{3}{9}$ equivalent to $\frac{1}{3}$? ______________

8. How do you know?

____

____

Name ____________________ Date ____________ Class ____________

LESSON 2-9

Puzzles, Twisters & Teasers

Are You in the Dark?

Fill in the boxes in the grids to show equivalent fractions. Then use the Answer Key to answer the riddle.

$\frac{S}{4}$ = $\frac{6}{8}$ = $\frac{9}{12}$

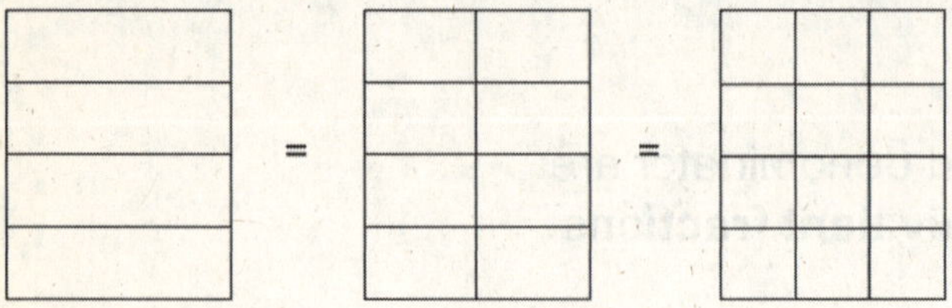

$\frac{G}{3}$ = $\frac{T}{6}$ = $\frac{4}{12}$

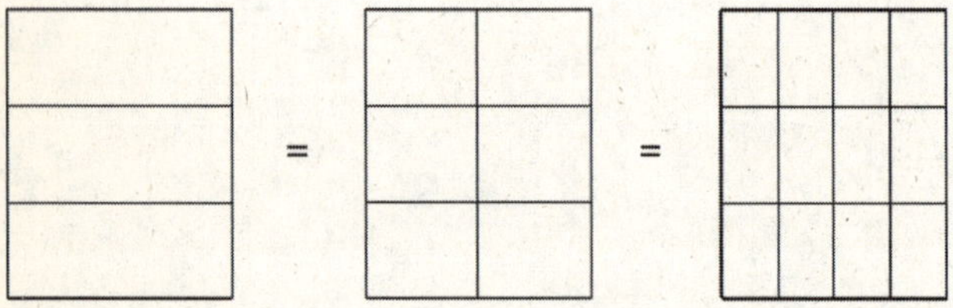

$\frac{3}{5}$ = $\frac{E}{10}$ = $\frac{D}{20}$

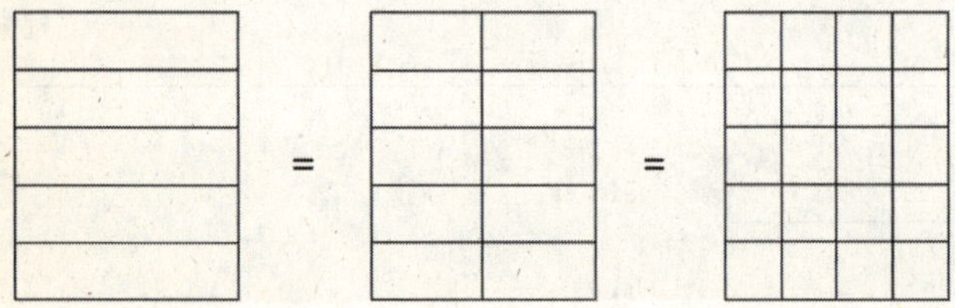

$\frac{2}{7}$ = $\frac{A}{14}$ = $\frac{R}{28}$

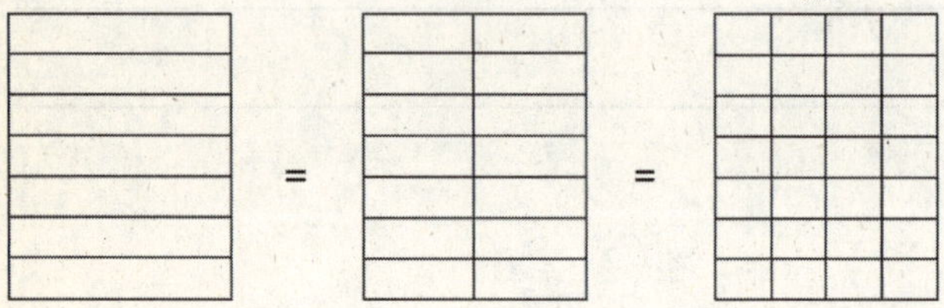

S = _____ G = _____ T = _____ E = _____ D = _____ A = _____ R = _____

Why did the tomato blush?

I ___ ___ ___ W ___ H ___
  2 3 4 2 6

___ ___ L ___ ___ ___ ___ ___ ___ ___ I N ___
3 4 4 12 12 8 6 3 3 1

Name ______________________ Date ______________ Class ______________

LESSON 2-10

Practice A

Equivalent Fractions and Decimals

Write each fraction as a decimal. Round to the nearest hundredth, if necessary.

1. $\frac{7}{10}$ ________ 2. $\frac{9}{20}$ ________ 3. $\frac{3}{4}$ ________ 4. $\frac{20}{25}$ ________

5. $\frac{3}{8}$ ________ 6. $\frac{7}{5}$ ________ 7. $\frac{21}{7}$ ________ 8. $\frac{5}{3}$ ________

9. $\frac{3}{10}$ ________ 10. $\frac{4}{5}$ ________ 11. $\frac{1}{25}$ ________ 12. $\frac{3}{20}$ ________

Write each decimal as a fraction or mixed number in simplest form.

13. 0.55 ________ 14. 0.03 ________ 15. –0.75 ________

16. 2.1 ________ 17. 5.25 ________ 18. 9.33 ________

19. 1.8 ________ 20. –1.74 ________ 21. 10.6 ________

22. –7.08 ________ 23. 0.625 ________ 24. 0.001 ________

Write each answer as a decimal rounded to the nearest thousandth.

25. Out of 45 times at bat, Raul got 19 hits. Find Raul's batting average.

____

26. On a test, Selena answered 26 out of 30 questions correctly. What portion of her answers was correct?

____

Name ____________________ Date ______________ Class ______________

LESSON 2-10

Practice B

Equivalent Fractions and Decimals

Write each fraction as a decimal. Round to the nearest thousandth, if necessary.

1. $\frac{2}{10}$ ________
2. $\frac{19}{20}$ ________
3. $\frac{5}{8}$ ________
4. $\frac{11}{5}$ ________
5. $\frac{19}{6}$ ________
6. $\frac{17}{4}$ ________
7. $\frac{13}{12}$ ________
8. $\frac{30}{7}$ ________
9. $\frac{7}{4}$ ________
10. $\frac{9}{20}$ ________
11. $\frac{11}{10}$ ________
12. $\frac{2}{25}$ ________

Write each decimal as a fraction in simplest form.

13. 0.85 ________
14. 0.11 ________
15. –0.25 ________
16. 4.3 ________
17. 7.75 ________
18. 5.03 ________
19. –1.06 ________
20. 0.375 ________
21. –2.65 ________
22. –5.6 ________
23. 1.12 ________
24. 0.005 ________

Write each answer as a decimal rounded to the nearest thousandth.

25. In the 1998 Winter Olympics, a total of 205 medals were awarded. The United States won 13 medals. What portion of the medals did the United States win?

____

26. On a test, Hailey answered 64 out of 75 question correctly. What portion of her answers was correct?

____

Name ____________________ Date ____________ Class ____________

LESSON 2-10

Practice C

Equivalent Fractions and Decimals

Write each fraction as a decimal. Round to the nearest thousandth, if necessary.

1. $\frac{5}{12}$ ________
2. $\frac{19}{4}$ ________
3. $\frac{25}{7}$ ________
4. $\frac{9}{8}$ ________
5. $\frac{17}{3}$ ________
6. $\frac{13}{11}$ ________
7. $\frac{15}{9}$ ________
8. $\frac{30}{13}$ ________
9. $\frac{9}{50}$ ________
10. $\frac{13}{25}$ ________
11. $\frac{23}{20}$ ________
12. $\frac{17}{4}$ ________

Write each decimal as a fraction in simplest form.

13. 0.25 ________
14. 0.13 ________
15. –0.65 ________
16. 4.8 ________
17. 1.875 ________
18. 9.06 ________
19. –1.008 ________
20. 0.625 ________
21. –2.05 ________

Determine if the numbers in each pair are equivalent.

22. $\frac{3}{7}$ and 0.41 ________
23. $\frac{13}{20}$ and 0.65 ________
24. $\frac{19}{25}$ and 0.78 ________

25. In the 1996 Summer Olympics, a total of 842 medals were awarded. The United States won 101 medals. What portion of the medals did the United States win? Write your answer as a decimal rounded to the nearest thousandth.

____

26. Pete Rose had 4,256 hits in 14,053 at bats. Write his batting average as a fraction and as a decimal rounded to the nearest thousandth.

____

Name ______________________ Date ______________ Class ______________

LESSON 2-10

Review for Mastery

Equivalent Fractions and Decimals

To write a fraction as a decimal, divide the numerator of the fraction by the denominator of the fraction.

Write $\frac{3}{7}$ as a decimal.

- Divide 3 by 7.
- To round your answer to the nearest hundredth, add 3 zeros after the decimal point in the divisor.

$$\begin{array}{r} 0.428 \\ 7\overline{)3.000} \\ -\underline{28}\downarrow \\ 20 \\ -\underline{14}\downarrow \\ 60 \\ -\underline{56} \\ 4 \end{array}$$

0.428 rounded to the nearest hundredth is 0.43.

1. Write $\frac{2}{5}$ as a decimal. $\frac{2}{5}$ = ____________

$$\begin{array}{r} 5\overline{)2.0} \\ -\underline{\quad} \\ \underline{\quad} \end{array}$$

Write each fraction as a decimal. Round to the nearest thousandth, if necessary.

2. $\frac{3}{4}$ ________ 3. $\frac{7}{8}$ ________ 4. $\frac{3}{2}$ ________ 5. $\frac{5}{3}$ ________

To write a decimal as a fraction:
Step 1: Use place value to read the decimal. Say the number aloud.
Step 2: Write a fraction for the number you just said.
Step 3: Simplify if necessary.

Write 0.005 as a fraction.
Read 0.005 as "five thousandths."

Write $\frac{5}{1000}$ for five thousandths.

Simplify: $\frac{5 \div 5}{1{,}000 \div 5} = \frac{1}{200}$

Write 1.6 as a fraction.
Read 1.6 as "one and six tenths."

Write $1\frac{6}{10}$ for one and six tenths.

Simplify: $1\frac{6 \div 2}{10 \div 2} = 1\frac{3}{5}$

Write each decimal as a fraction or mixed number in simplest form.

6. 0.8 ________ 7. 2.25 ________ 8. –0.02 ________

Name ________________ Date ________ Class ________

LESSON 2-10

Challenge

Decimal Patterns

Repeating decimals continue without end with a repeating pattern. Some decimals continue without end, but do not have a repeating pattern. These decimals are called *nonrepeating, nonterminating* decimals. Although they continue without end, these decimals may have a pattern.

Examples of nonrepeating, nonterminating decimals with patterns:

5.121121112…

0.881882883…

Write *r* if the decimal is a repeating decimal and *n* if the decimal is a nonrepeating, nonterminating decimal.

1. 4.562222… ________
2. 6.1323323332… ________
3. 8.355355… ________
4. 0.230230023… ________
5. 10.7727227… ________
6. 4.121314… ________
7. 0.42818181… ________
8. 29.7053053… ________
9. 1.221221222… ________
10. 5.67167672… ________
11. 75.320333… ________
12. 4.121231234… ________

Identify the pattern in each nonrepeating, nonterminating decimal. Use the pattern to extend the decimal several places.

13. 0.010203… ________
14. 8.636336333… ________
15. 25.121231234… ________
16. 9.989796… ________
17. Write an example of a repeating decimal.

____

18. Write an example of a nonrepeating, nonterminating decimal with a pattern.

____

Name ________________ Date ________________ Class ________________

LESSON 2-10

Problem Solving

Equivalent Fractions and Decimals

Write the correct answer.

1. On a test, Shane answered 37 out of 40 questions correctly. What portion of his answers was correct? Write your answer as a decimal rounded to the nearest thousandth.

____

2. Sammy Sosa hit 66 home runs in 1998. He had 643 at bats. Write his home run average as a decimal rounded to the nearest thousandth.

____

3. In February, Chicago receives an average of $1\frac{2}{5}$ inches of rain. Write a decimal to show the number of inches of rain.

____

4. On a test, Ellen answered 51 out of 64 questions correctly. What portion of her answers was correct? Write your answer as a decimal rounded to the nearest thousandth.

____

Choose the letter for the best answer.

Use the graph for 5–6.

World Energy Consumption 2003
(in quadrillions of British Thermal Units, BTU)

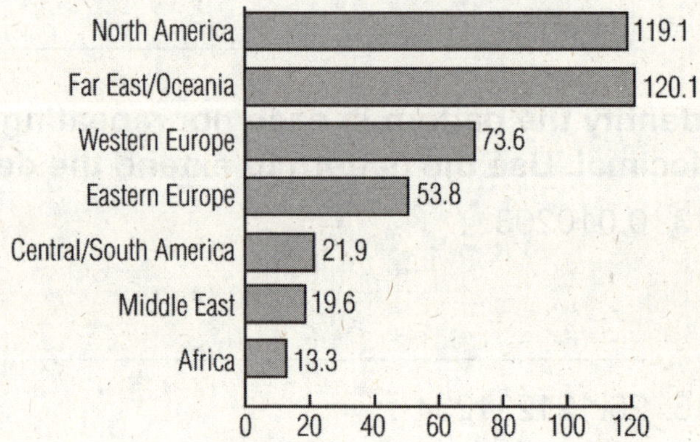

5. Which mixed number shows the energy, in quadrillions of BTU, consumed in Eastern Europe in 2003?

A 53 C $53\frac{1}{2}$

B $53\frac{1}{4}$ D $53\frac{4}{5}$

6. Which mixed number shows the energy, in quadrillions of BTU, consumed in the Middle East in 2003?

F $19\frac{1}{10}$ H $19\frac{3}{5}$

G $19\frac{2}{5}$ J $19\frac{3}{10}$

7. Jill sold 478 out of 520 tickets to the opening night of her theater performance. What portion of the tickets did she sell?

A 0.998 C 0.919

B 1.088 D 0.081

8. The high school sold 369 out of 460 tickets to the opening night of a concert series. What portion of the tickets was sold?

F 0.198 H 1.247

G 0.829 J 0.802

Name ______________________ Date ______________ Class ______________

LESSON 2-10

Reading Strategies

Compare and Contrast

Compare what happens when fractions are changed to decimals.

$\frac{2}{5}$ • Read $\frac{2}{5}$ as "2 divided by 5." • Write ⟶ $2 \div 5$

Change a fraction to a decimal by dividing the numerator by the denominator.

$5\overline{)2.0}$

$-2\,0$

$\frac{2}{5} = 0.4$ The dividing ends, or terminates, with no remainder. 0.4 is called a terminating decimal.

1. Is there a remainder in the problem? How do you know?

____

____

2. What do we call a decimal that ends with no remainder?

____

$\frac{2}{6}$ • Read $\frac{2}{6}$ as "2 divided by 6." • Write ⟶ $2 \div 6$

−18
20
−18
2

$\frac{2}{6} = 0.333\ldots$ or $0.\overline{3}$

← Note how dividing continues in a pattern. The number 0.333 . . . is a repeating decimal. The bar over the 3 means 3 repeats.

Answer each question.

3. Compare the division of $\frac{2}{5}$ to the division of $\frac{2}{6}$. What is the difference?

____

____

4. What is the name for a decimal with a remainder that has a repeating pattern?

____

Name ______________________ Date ____________ Class ____________

LESSON 2-10

Puzzles, Twisters & Teasers

The Terminator!

Find the words on the list in the word search. Four of the letters can be used to answer the riddle. (Hint: the names of the individual letters sound like the riddle's answer.)

fraction	decimal	terminating	repeating	average	pick
statistic	convert	numerator	divide	pattern	denominator

```
D E N O M I N A T O R U J T K P
I E R S T A T I S T I C F E T A
V W C A V E R A G E N O D R K T
I F V I Z X C V B Y H N U M W T
D R Q E M T G B Y H A V L I Z E
E A S D U A B G R V I E C N I R
P C O L C O L P I C K R O A D N
N U M E R A T O R A U T G T E I
I I W E R P A R E P E A T I N G
U O Z A Q W S X C O T G H N J I
Y N G F R A C T I O N A T G O R
```

What will the dentist say to you if you do not brush your teeth?

_____ _____ _____ _____

Name ______________________ Date ______________ Class ______________

LESSON 2-11

Practice A

Comparing and Ordering Rational Numbers

Compare the fractions. Write < or >.

1. $\frac{4}{9}$ ____ $\frac{7}{9}$

2. $-\frac{5}{6}$ ____ $-\frac{1}{6}$

3. $-\frac{2}{3}$ ____ $-\frac{3}{4}$

4. $-\frac{1}{4}$ ____ $-\frac{1}{7}$

5. $1\frac{5}{8}$ ____ $1\frac{9}{10}$

6. $3\frac{1}{5}$ ____ $-3\frac{2}{5}$

Compare the decimals. Write < or >.

7. 0.053 ____ 0.312

8. –2.51 ____ –2.511

9. 1.450 ____ 1.405

10. $1.\overline{18}$ ____ 1.188

11. –5.21 ____ 5.00

12. $9.\overline{03}$ ____ 9.303

Write the numbers in order from least to greatest.

13. $0.\overline{5}$, 0.5, $\frac{5}{8}$

____

14. 1.34, $1\frac{1}{3}$, 1.3

____

15. 2.07, $2\frac{7}{10}$, 2.67

____

16. –4.8, –4.08, $-4\frac{1}{8}$

____

17. 5.78, $-5\frac{7}{8}$, –5.9

____

18. $\frac{6}{7}$, $\frac{5}{6}$, 0.8

____

19. $2\frac{2}{3}$, 2.65, 2.69

____

20. –0.38, $-\frac{3}{8}$, –0.04

____

21. $\frac{3}{7}$, 0.45, $\frac{4}{9}$

____

22. Twenty-four karat gold is considered pure. If a ring is 14 karat gold, what is its purity as a fraction and as a decimal rounded to the nearest thousandth?

____

23. Amy scored 9.02 on her balance beam routine. Joella scored 9.2 on her balance beam routine. Which gymnast received the higher score?

____

Name ______________________ Date ______________ Class ______________

LESSON 2-11

Practice B

Comparing and Ordering Rational Numbers

Compare the fractions. Write < or >.

1. $-\frac{7}{8}$ ____ $-\frac{5}{8}$

2. $\frac{3}{10}$ ____ $\frac{3}{8}$

3. $4\frac{7}{12}$ ____ $5\frac{5}{12}$

4. $\frac{35}{40}$ ____ $\frac{22}{50}$

5. $-\frac{1}{2}$ ____ $\frac{3}{4}$

6. $-3\frac{7}{8}$ ____ $-3\frac{9}{10}$

Compare the decimals. Write < or >.

7. –0.531 ____ –0.513

8. 0.73 ____ 0.073

9. $3.\overline{59}$ ____ 3.599

10. 0.121 ____ 0.12

11. –7.56 ____ –5.76

12. –.099 ____ –.097

Order the numbers from least to greatest.

13. $\frac{4}{9}$, 0.4, 0.45

____

14. 1.7, 1.65, $1\frac{2}{3}$

____

15. 3.18, $3\frac{1}{8}$, 3.80

____

16. –5, –5.25, $-5\frac{2}{5}$

____

17. $-6\frac{3}{4}$, 6.34, –6.4

____

18. $\frac{11}{12}$, $\frac{8}{9}$, 0.91

____

19. $-\frac{3}{5}$, $-\frac{5}{7}$, –0.65

____

20. 0.3, 0.345, $\frac{1}{3}$

____

21. –0.75, $\frac{7}{8}$, $-\frac{5}{8}$

____

22. A ream of paper contains 500 sheets of paper. Norm has 373 sheets of paper left from a ream. Express the portion of a ream Norm has as a fraction and as a decimal.

____

23. The density of Venus, compared to Earth having a density of 1, is 0.943. The density of Mercury is 0.983, compared to the density of Earth. Which planet has a greater density, Venus or Mercury?

____

Name ______________________ Date ______________ Class ______________

LESSON 2-11

Practice C

Comparing and Ordering Rational Numbers

Compare the fractions. Write < or >. Justify your answer.

1. $-4\frac{3}{4}$ ____ $-4\frac{5}{6}$
2. $\frac{9}{10}$ ____ $\frac{10}{11}$
3. $2\frac{2}{5}$ ____ $2\frac{3}{8}$
4. $-12\frac{6}{7}$ ____ $-12\frac{8}{11}$
5. $-\frac{3}{4}$ ____ $\frac{6}{9}$
6. $-\frac{99}{100}$ ____ $-\frac{900}{1000}$

Compare the decimals. Write < or >. Justify your answer.

7. 1.031 ____ 1.3
8. –4.25 ____ –4.5
9. $2.\overline{17}$ ____ 2.17
10. –1.52 ____ –1.025
11. $22.\overline{31}$ ____ $22.\overline{32}$
12. $-0.1\overline{893}$ ____ –0.1899

Order the numbers from least to greatest.

13. $\frac{9}{16}$, 0.54, 0.65

____

14. 3.49, 3.14, $3\frac{4}{9}$

____

15. –3.05, –3.5, $-3\frac{1}{10}$

____

16. $-2\frac{7}{12}$, 2.59, –2.7

____

Choose the greater number.

17. $\frac{4}{5}$ or $\frac{5}{6}$ ____________
18. 0.095 or –0.9 ____________
19. $-\frac{1}{4}$ or –0.025 ____________
20. –5 or –5.01 ____________

21. A horsepower is the power needed to lift 33,000 pounds 1 foot in 1 minute. It is equivalent to 746 watts. Express 525 watts as a portion of a horsepower both as a fraction and a decimal rounded to the nearest thousandth.

____

22. The density of Jupiter, compared to Earth having a density of 1, is 0.24. The density of Saturn is 0.124, compared to the density of Earth. Which planet has a greater density, Jupiter or Saturn?

____

Name ____________________ Date ____________ Class ____________

LESSON 2-11

Review for Mastery

Comparing and Ordering Rational Numbers

You can use fraction strips to compare two fractions that have different denominators.

Compare $\frac{2}{3}$ and $\frac{1}{2}$.

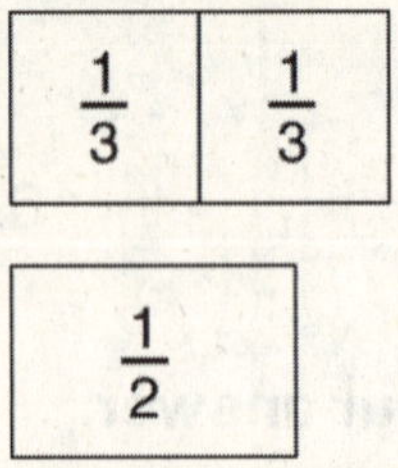

$\frac{2}{3}$ is greater than $\frac{1}{2}$.

$\frac{2}{3} > \frac{1}{2}$

Compare $\frac{2}{5}$ and $\frac{3}{4}$.

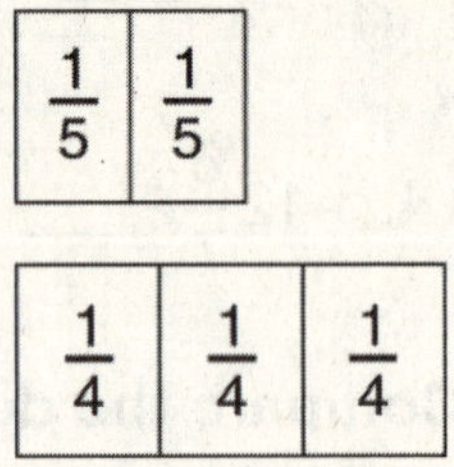

$\frac{2}{5}$ is less than $\frac{3}{4}$.

$\frac{2}{5} < \frac{3}{4}$

Use fraction strips to compare. Write < or >.

1. $\frac{1}{3}$ ____ $\frac{1}{2}$

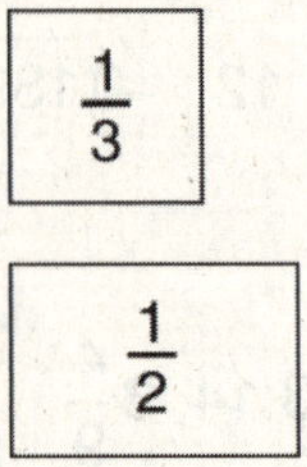

2. $\frac{3}{4}$ ____ $\frac{5}{8}$

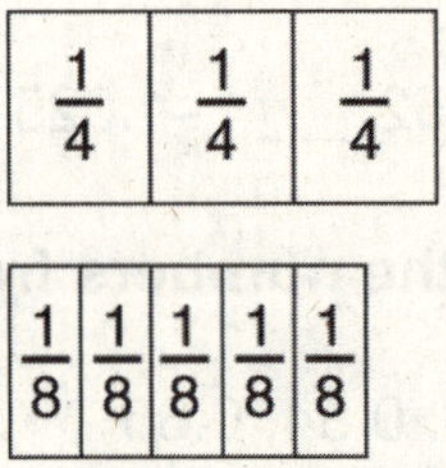

3. $\frac{1}{2}$ ____ $\frac{3}{5}$

4. $\frac{1}{5}$ ____ $\frac{1}{8}$

5. $\frac{3}{8}$ ____ $\frac{1}{3}$

To compare decimals:

1) Add zeros so that each decimal has the same number of decimal places.
2) Ignore the decimal points and compare the numbers as if they were integers.

Compare 4.3 and 4.27.

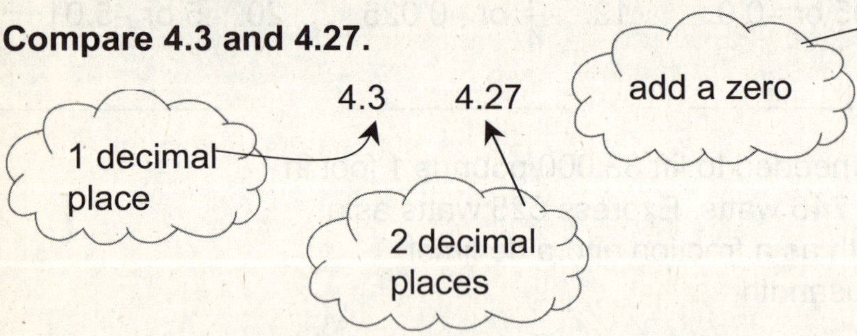

4.30 4.27

430 > 427, so 4.3 > 4.27.

Compare the decimals using < or >.

6. 7.2 ____ 7.8

7. 0.64 ____ 0.604

8. 2.09 ____ 2.817

Name ______________________ Date ______________ Class ______________

LESSON 2-11

Challenge

Another Rational Number

Between any two rational numbers, there are other rational numbers.

Examples: $\frac{2}{3}$ lies between $\frac{1}{2}$ and $\frac{5}{6}$.

–6 lies between –5.9 and –6.1.

$1\frac{1}{4}$ lies between 1 and $1\frac{1}{2}$.

Write a rational number that lies between each pair of rational numbers below.

1. 0 and $\frac{1}{5}$ ______________
2. 3.4 and 3.5 ______________
3. 1.08 and 1.8 ______________
4. –7 and –7.5 ______________
5. –12.1 and –12.2 ______________
6. 0 and $-\frac{1}{2}$ ______________
7. $1\frac{1}{8}$ and $1\frac{1}{2}$ ______________
8. $-2\frac{1}{10}$ and $-2\frac{1}{2}$ ______________
9. 3.01 and 3.02 ______________
10. 4.2 and $4\frac{1}{2}$ ______________
11. –1 and –1.1 ______________
12. 3.5 and $3\frac{7}{8}$ ______________
13. $-2\frac{3}{8}$ and –2.8 ______________
14. 0 and –0.3 ______________
15. $7\frac{1}{10}$ and 7.5 ______________
16. $5\frac{7}{8}$ and $5\frac{11}{12}$ ______________
17. $-\frac{2}{3}$ and $-\frac{3}{4}$ ______________
18. $3\frac{1}{5}$ and 3.19 ______________

Name ______________________ Date ______________ Class ______________

LESSON 2-11

Problem Solving

Comparing and Ordering Rational Numbers

Write the correct answer.

1. Some of the wettest summers in the Northeast had the following amounts of precipitation, in inches: $14\frac{7}{10}$, $14\frac{1}{4}$, $14\frac{4}{5}$, and 14.69. Order the amounts of precipitation from least to greatest.

____

____

2. The U.S. soybean crop decreased each year from 2001 to 2003. Production for each year was approximately $2\frac{3}{4}$, 2.9, $2\frac{2}{5}$ billion bushels. Match the soybean crop, in billions of bushels, with its year.

____

____

3. Mark's batting average last year was 0.334. Ken's was 0.304, and Ty's was 0.343. Write the batting averages in order from lowest to highest.

____

4. Kiki lives 0.097 miles from the school, and George lives 0.131 miles from the school. Who lives farthest from the school?

____

Choose the letter for the best answer. Use the graph for problems 5–6.

5. In which year did NFL teams have the greatest success rate?

A 1992 C 1996

B 1994 D 2000

6. In which year was the success rate of NFL teams greater than 0.775 and less than 0.78?

F 1994 H 1997

G 1995 J 1999

NFL Field Goal Success Rate

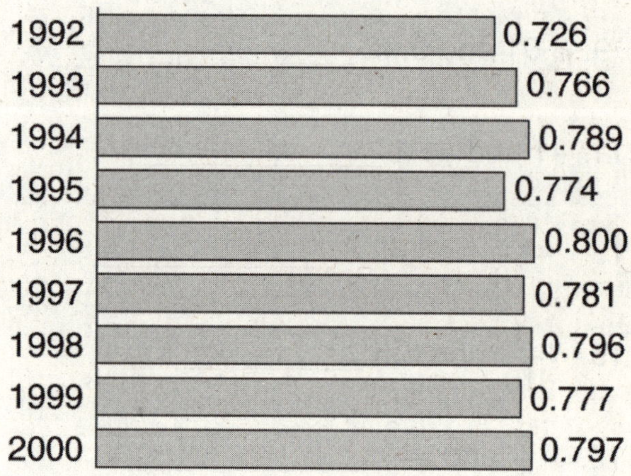

Name ______________________ Date ______________ Class ______________

LESSON 2-11

Reading Strategies

Compare Visually

You can compare fractions that have the same **denominator** by comparing their **numerators.**

Which is greater, $\frac{4}{9}$ or $\frac{5}{9}$?

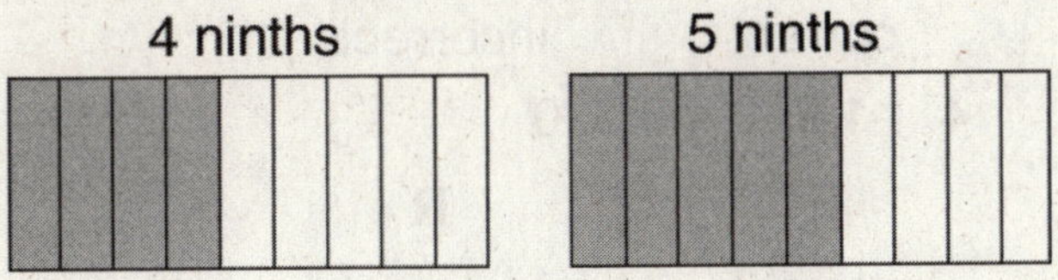

Notice that both fractions are naming ninths.

Look to see which fraction has the most ninths. ⟶ $\frac{5}{9} > \frac{4}{9}$

You can compare fractions visually by using fraction strips.

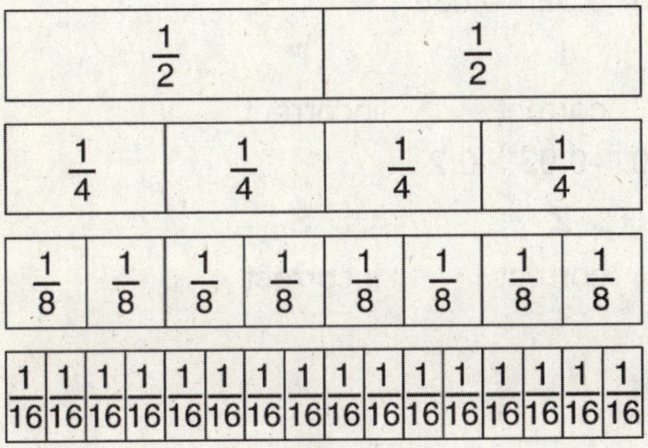

Compare the following fractions. Use the fraction strips above to help you compare. Write >, <, or =.

1. $\frac{1}{2}$ ____ $\frac{1}{8}$

2. $\frac{2}{16}$ ____ $\frac{3}{8}$

3. $\frac{3}{4}$ ____ $\frac{7}{8}$

4. $\frac{4}{16}$ ____ $\frac{2}{8}$

5. $\frac{8}{16}$ ____ $\frac{2}{4}$

6. $\frac{2}{4}$ ____ $\frac{2}{16}$

Name ______________________ Date ______________ Class ______________

LESSON 2-11

Puzzles, Twisters & Teasers

What's the Big Deal?

Decide whether or not each equation is correct. Circle the letters above your answers. Then solve the riddle.

1. 3.8 < 3.6

   B correct **O** incorrect

2. 0.088 > 0.109

   U correct **T** incorrect

3. −4.26 > −4.266

   H correct **L** incorrect

4. −1.902 < 0.920

   E correct **R** incorrect

5. 0.755 < 0.7

   N correct **M** incorrect

6. 1.82 > 1.6

   A correct **H** incorrect

7. 2.88 < −2.98

   O correct **T** incorrect

8. 0.32 > 0.088

   I correct **P** incorrect

9. 0.999 < 1.0

   C correct **F** incorrect

10. −0.93 > 0.2

    Z correct **S** incorrect

What is an insect's favorite subject in school?

M ___ ___ ___ ___ ___ ___ ___

Answers

LESSON 2-1

Practice A

1. 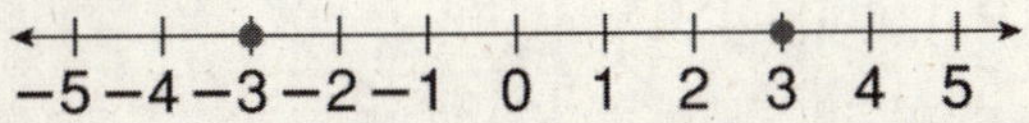

2. 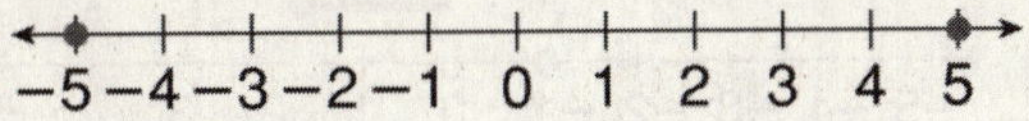

3. < 4. >

5. > 6. <

7. 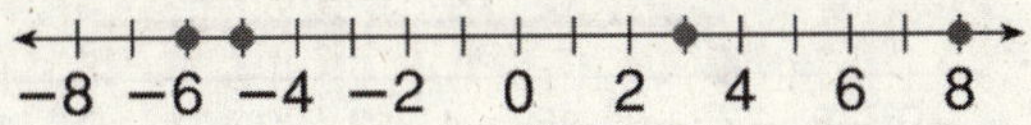

–6; –5; 3; 8

8. 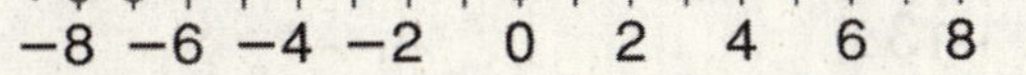

–8; –7; 0; 6

9. 6 10. 2
11. 1 12. 8
13. 9 14. 3
15. 4 16. 10
17. 15 18. 20
19. 13 20. 17
21. –5 22. 15

Practice B

1. 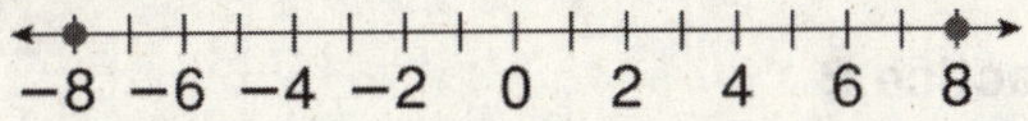

2. 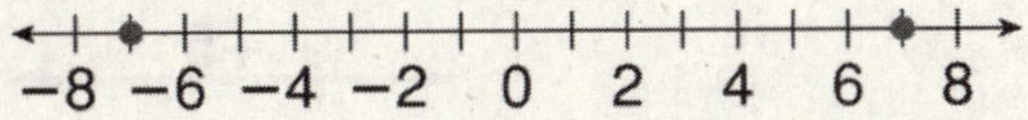

3. < 4. >

5. < 6. >

7. 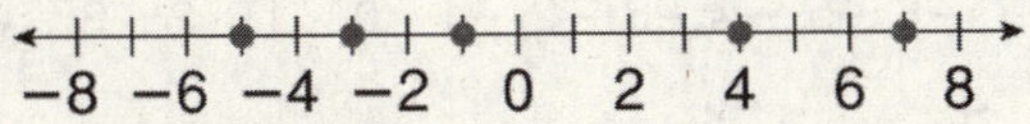

–5; –3; –1; 4; 7

8. 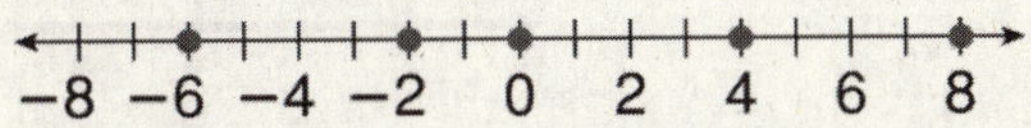

–6; –2; 0; 4; 8

9. 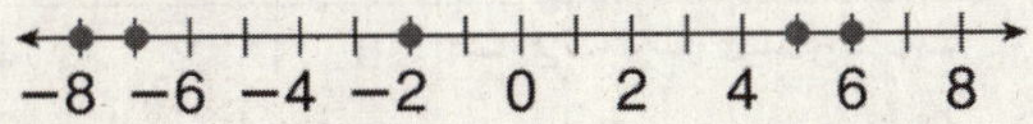

–8; –7; –2; 5; 6

10. 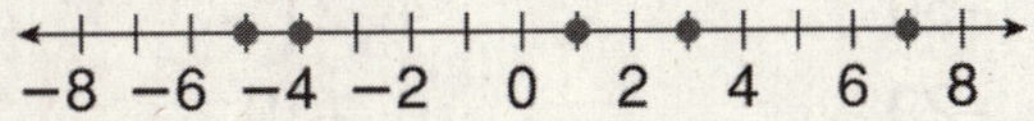

–5; –4; 1; 3; 7

11. 18 12. 11
13. 25 14. 19
15. 10 16. 16
17. 22 18. 14
19. 9 20. 24
21. 7 22. 17
23. –12 24. 6,684

Practice C

1. 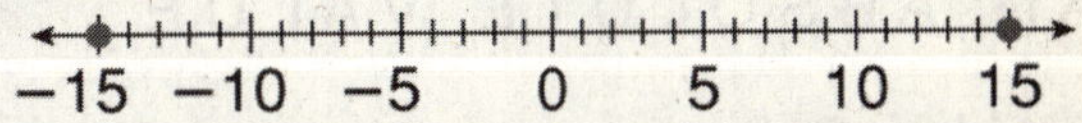

2. 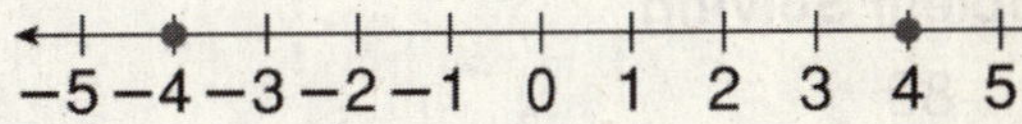

3. < 4. >

5. > 6. <

7. 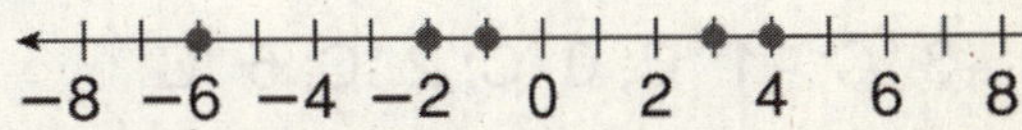

–6; –2; –1; 3; 4

8. 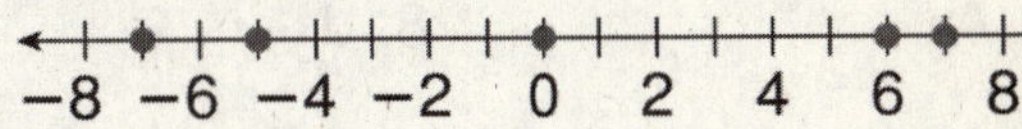

–7; –5; 0; 6; 7

9. 45 10. 38
11. 99 12. 22
13. 19 14. 375
15. 59 16. 84
17. = 18. <
19. < 20. <
21. > 22. <
23. > 24. <
25. 750,000; –2,000,000
26. –30; 110

Review for Mastery

1. 4 2. 4
3. opposite integers

4. 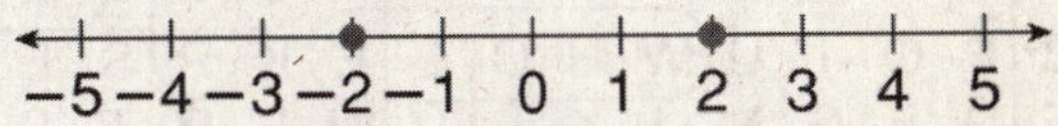

5. 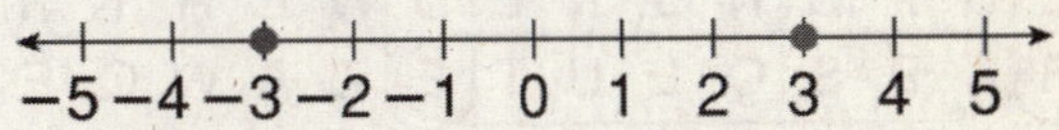

6. –3; –1; 2 7. –6; –2; 4
8. –3; 1; 7 9. 3

Challenge

1. –12 < –11 < 6
2. –201 < –3 < –1
3. –6 < | –3 | < | –6 |
4. –7 < | 3 | < | –38 |
5. 4 > –4 > –40
6. | –18 | > 0 > –8
7. | –75 | > | 12 | > 7
8. 1 > –12 > –16

A N A B S O L U T E V A L U E

Problem Solving

1. –80
2. outside Jared's house
3. –62,500; 34,100
4. –3 °C; –1 °C; 0 °C; 2 °C; 5 °C
5. C
6. F
7. A
8. J

Reading Strategies

1. the set of whole numbers and their opposites
2. –6
3. –10
4. 0
5. 8
6. 3
7. integer
8. not an integer
9. not an integer
10. integer

Puzzles; Twisters & Teasers

M	H	T	E	F	V	C	D	E	S	W	I	C	F	R
O	P	P	O	S	I	T	E	V	F	R	N	B	H	U
X	Z	O	I	E	S	W	A	X	N	M	T	A	W	H
N	E	G	A	T	I	V	E	C	P	U	E	I	H	Q
U	R	A	S	D	F	G	H	V	L	R	G	P	O	K
M	O	Z	X	C	V	B	N	A	R	J	E	L	L	Y
B	P	O	I	U	W	H	O	L	E	K	R	S	E	M
E	D	F	G	H	J	K	L	U	W	E	R	T	S	X
R	A	B	S	O	L	U	T	E	Z	A	W	Q	E	C

J E L L Y

LESSON 2-2

Practice A

1.

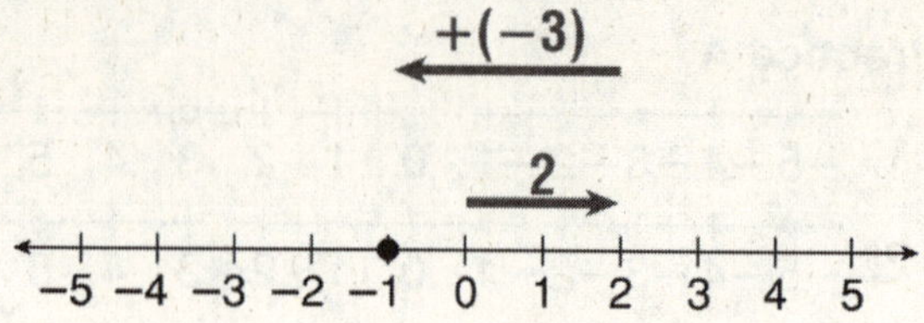

–1

2.

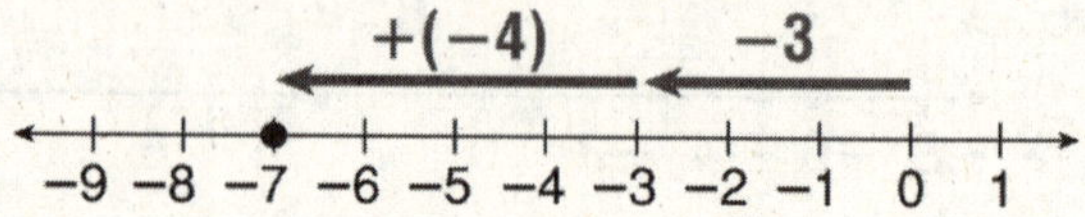

–7

3. –13
4. –1
5. –1
6. –3
7. 12
8. 4
9. 8
10. –16
11. –5
12. –10
13. 5
14. 4
15. 7
16. –10
17. 5
18. –1
19. 2
20. –5
21. 11 °F
22. 3 yards

Practice B

1.

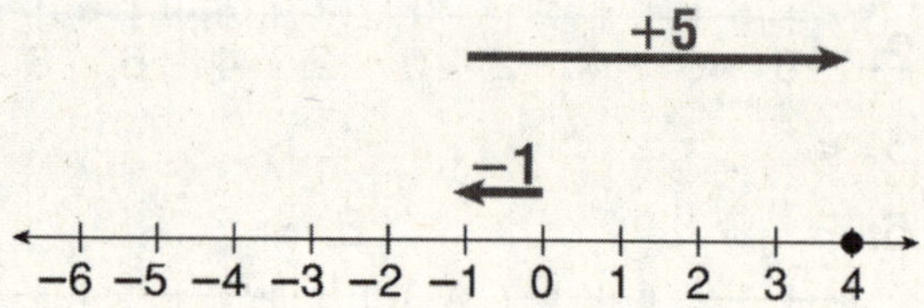

4

2.

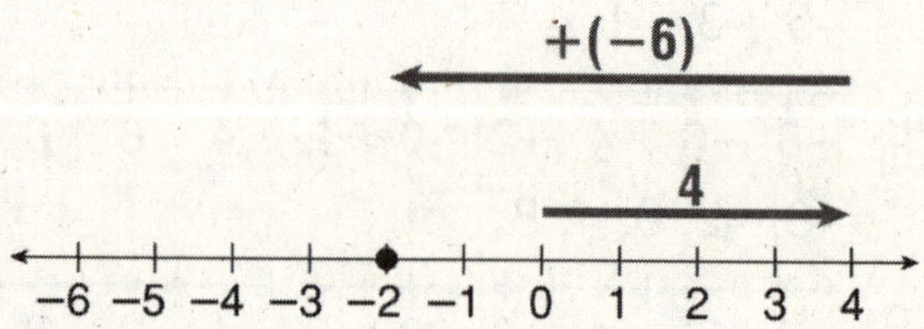

–2

3. –60
4. 21
5. –29
6. –10
7. 43
8. –34
9. –14
10. –40

11. –6
12. 12
13. –22
14. –2
15. –15
16. –24
17. 13
18. –30
19. 0
20. –18
21. 4 °F
22. –9 points

Practice C

1.
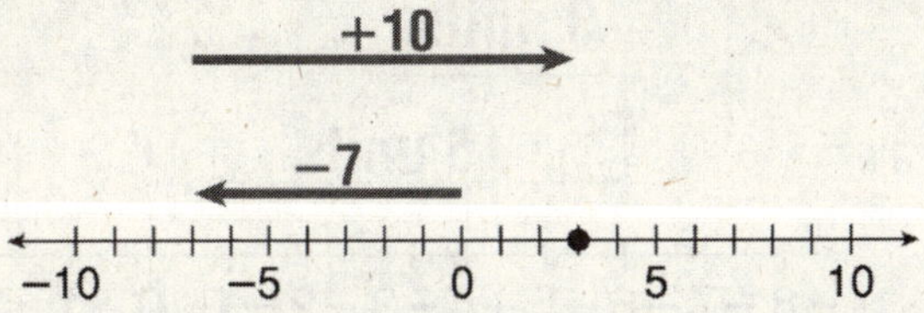

3

2.
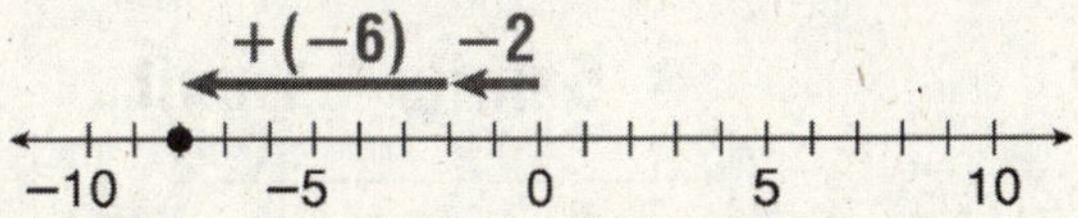

–8

3. –36
4. –23
5. 29
6. 70
7. 19
8. –94
9. –14
10. –96
11. 6
12. –2
13. –23
14. –4
15. 38
16. –80
17. –29
18. =
19. >
20. <
21. –1 °F
22. –6 points or a loss of 6 points

Review for Mastery

1. subtract
2. negative
3. add
4. negative
5. –13
6. 4
7. –5
8. –1
9. –4
10. 2
11. –5

Challenge

1. –13
2. –22
3. –22
4. –27
5. –2
6. –16
7. –19
8. –5
9. –18
10. >
11. =
12. >
13. >
14. <
15. >
16. <
17. <
18. <
19. Answers will vary.
20. Possible answer: chalk
21. Possible answer: bench
22. Answers will vary.

Problem Solving

1. 5 °F
2. –22 °F
3. $35
4. 535 ft
5. –97 ft
6. 17,500 ft
7. B
8. H
9. C
10. J

Reading Strategies

1. 0
2. to the right; 6
3. to the left; 4
4. 0
5. to the left; 5
6. to the left; 3

Puzzles, Twisters & Teasers

R: –9
S: 7
L: –19
H: –88
O: –6
E: –18
C: 39
W: 25
D: 13
A: –17
T: –16
N: 1

S H E W A N T E D
C O L D H A R D C A S H

LESSON 2-3

Practice A

1. 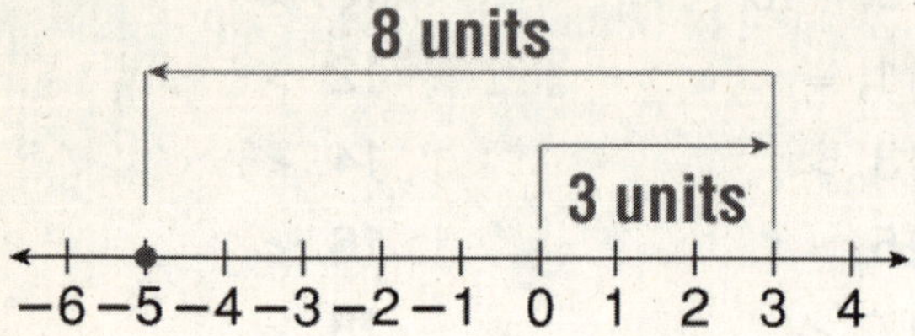

−5

2. 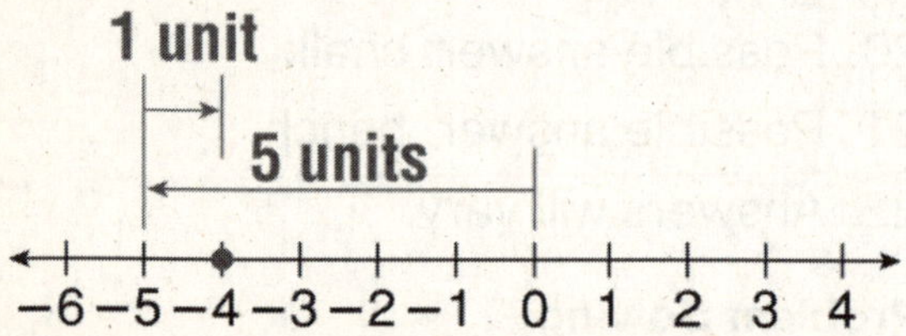

−4

3. −7
4. −5
5. 6
6. 9
7. −16
8. 0
9. 1
10. 12
11. 7
12. 16
13. −11
14. 4
15. 9
16. −8
17. 3
18. −2
19. 0
20. −10
21. 9 °F
22. 14 °F

Practice B

1. 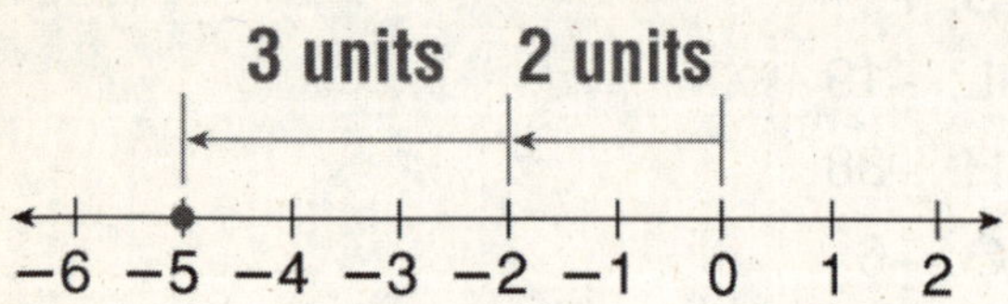

−5

2. 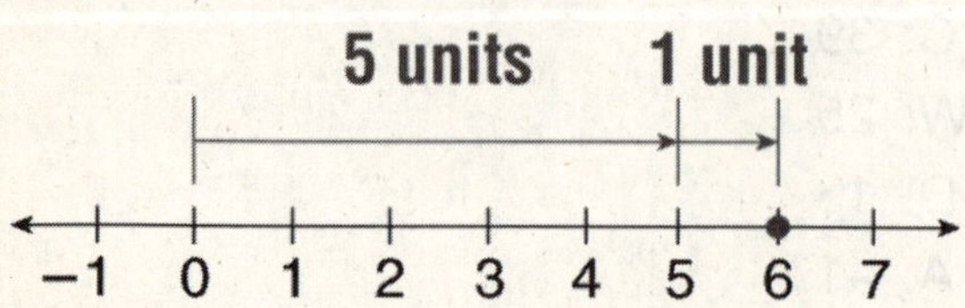

6

3. −10
4. 5
5. −4
6. 24
7. 0
8. 46
9. −1
10. 42
11. −6
12. −26
13. 30
14. −5
15. 16
16. −22
17. 7
18. 0
19. 29
20. −5
21. 4 °F
22. 7 °F

Practice C

1. 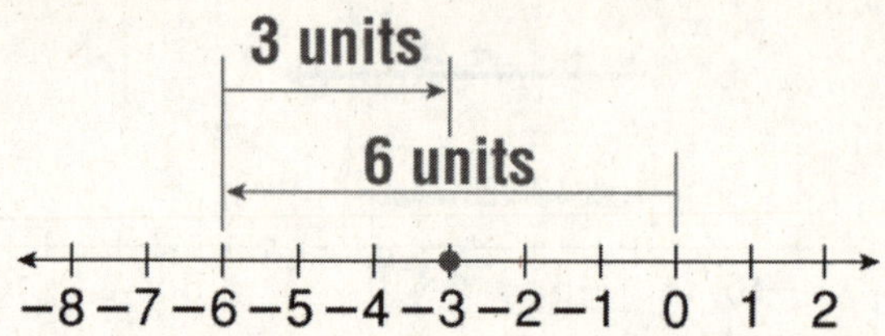

−3

2. 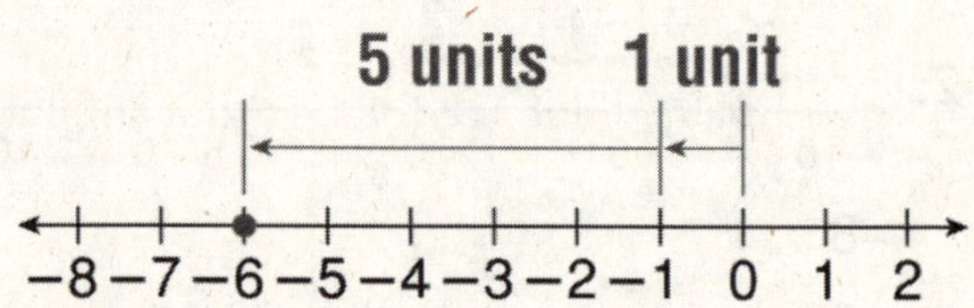

−6

3. −31
4. −12
5. −16
6. 38
7. 66
8. −74
9. 0
10. −52
11. −48
12. 10
13. −7
14. −12
15. 14
16. 0
17. −20
18. 25
19. −57
20. 144°F
21. −5 °F

Review for Mastery

1. a. 5; b. −1; c. 20
2. a. negative; b. 2; c. −2
3. a. positive; b. 8; c. 8
4. 40
5. −3
6. −26
7. 0
8. 31
9. −5

Challenge

Temperature Facts	°F
Louisiana's lowest	–16°
Texas's highest	120°
Fairbanks normal January	–10°
Tampa normal July	82°
Lowest in Detroit in 2000	–3°
Highest in Norfolk in 2000	96°
North Carolina's lowest	–34°
South Carolina's lowest	–19°

Highest and Lowest Points on Continents	Elevation
Death Valley, North America	–282 ft
Caspian Sea, Europe	–92 ft
Mount Everest, Asia	29,035 ft
Mount McKinley, North America	20,320 ft
Lake Eyre, Australia	–52 ft
Dead Sea, Asia	–1,348 ft
Mount Elbrus, Europe	18,510 ft
Lake Assal, Africa	–512 ft

Problem Solving

1. 610 °C
2. $35,000
3. $118,000
4. 240 °C
5. B
6. J
7. D
8. F

Reading Strategies

1. to the left
2. 7
3. to the right
4. 3
5. –4
6. to the right; 2
7. to the left; 6
8. 4

Puzzles, Twisters & Teasers

C: –10
A: –9
J: 5
N: –4
B: –46
U: 18
O: –30
I: 50
G: –17
D: –25
Y: 8
E: –3
R: –7
T: –14
F: –6

A N U N I D E N T I F I E D
F R Y I N G O B J E C T

LESSON 2-4

Practice A

1. –6
2. –8
3. 12
4. –16
5. –35
6. –63
7. 32
8. 15
9. 25
10. –32
11. 42
12. –72
13. –7
14. 20
15. –18
16. 49
17. –3
18. 5
19. –4
20. 3
21. 9
22. –2
23. –6
24. 8
25. 7
26. 9
27. –4
28. –7
29. –6
30. 5
31. 3
32. undefined
33. –3 feet
34. 5 hours

Practice B

1. –40
2. –28
3. 18
4. –8
5. –36
6. –45
7. 48
8. 21
9. 36
10. –27
11. 16
12. –35
13. 80
14. –45
15. –54
16. 44
17. –5
18. 9
19. –2
20. 7
21. 6
22. undefined
23. –6
24. 9
25. –4
26. –5
27. 9
28. 5
29. –7
30. 8
31. 5
32. undefined
33. –2 feet
34. –3 degrees

Practice C

1. –72
2. –49
3. 30
4. –32
5. –120
6. –150
7. 80
8. 120
9. 35
10. –90
11. 64
12. –36
13. –72
14. –60
15. 125
16. –300
17. –9
18. 12
19. –4
20. 9
21. 8
22. undefined
23. –8
24. 5
25. –11
26. –5
27. 25
28. 12
29. –4
30. 20
31. 4
32. undefined
33. –5 feet
34. –$60

Review for Mastery

1. –20
2. –16
3. 1
4. –18
5. –21
6. 32
7. –30
8. 81
9. –9
10. –3
11. 4
12. –6
13. 3
14. –4
15. –5
16. 4

Challenge

Phil	Sam
85 yd	106 yd
–1,282 yd	–1,482 yd
–1,197 yd	–1,376 yd
–74.8 yd	–86 yd
	✓

Problem Solving

1. –2 °F
2. 1 °F
3. 3 °F
4. –1 °F
5. C
6. H
7. C
8. H

Reading Strategies

1. having $10
2. owing $17
3. to the left
4. 4
5. to the left
6. 4
7. to the left
8. 4
9. $12
10. $16

Puzzles, Twisters & Teasers

H: –9

E: 8

N: –8

T: –12

I: 9

S: –6

C: –4

G: 6

A: 12

R: 10

WHEN THERE IS CHANGE IN THE WEATHER

LESSON 2-5

Practice A

1. $n = 4$
2. $x = -3$
3. $a = 12$
4. $y = -2$
5. $c = -10$
6. $v = -1$
7. $j = -2$
8. $k = -8$
9. $s = 5$
10. $m = 10$
11. $d = -18$
12. $r = -28$
13. $p = -14$
14. $b = -3$
15. $f = 8$
16. $n = -24$
17. $k = -7$
18. $a = -4$
19. $x = 6$
20. $e = 9$
21. $m = 6$
22. 4 °F
23. $20

Practice B

1. $y = 1$
2. $n = -5$
3. $x = 28$
4. $p = -4$
5. $q = -8$
6. $w = -4$
7. $h = -4$
8. $b = -12$
9. $u = 25$
10. $d = -35$
11. $c = -80$
12. $s = -81$
13. $f = -31$
14. $v = -25$
15. $g = -2$
16. $s = 7$
17. $m = -24$
18. $w = 48$
19. $x = -16$
20. $e = -56$
21. $n = 27$
22. −44°F
23. $10

Practice C

1. $h = -2$
2. $a = 17$
3. $g = 33$
4. $m = -17$
5. $n = -30$
6. $z = -50$
7. $y = -5$
8. $d = 0$
9. $v = 18$
10. $k = 70$
11. $r = -1{,}000$
12. $t = -121$
13. $s = -43$
14. $j = 24$
15. $x = -8$
16. $n - 8 = -13$; $n = -5$
17. $-12f = -72$; $f = 6$
18. $-47 + d = 84$; $d = 131$
19. $8q = -40$; $q = -5$
20. $12.50
21. cost: $23.50; selling price: $29.50

Review for Mastery

1. 9; 9
   $p = 6$
2. 2; 2
   $w = -12$
3. 12; 12
   $x = 7$
4. 8; 8
   $f = 14$
5. $m = 13$
6. $s = 6$
7. $y = -6$
8. $a = -12$
9. $b = -24$
10. $t = 6$
11. $x = -12$
12. $k = -13$
13. $j = 13$
14. 5; 5; $g = -7$
15. −8; −8; $y = 12$
16. −6; −6; $f = -9$
17. $e = -11$
18. $n = -7$
19. $c = 15$
20. $n = -60$
21. $m = 54$
22. $s = -80$
23. $w = -24$
24. $z = 45$
25. $h = -66$

Challenge

1. C; $y = -12$
2. G; $x = -17$
3. D; $x = 5$
4. B; $n = -32$
5. F; $y = 50$
6. J; $y = 16$
7. H; $n = 8$
8. E; $n = -7$
9. A; $x = 4$
10. I; $y = 3$

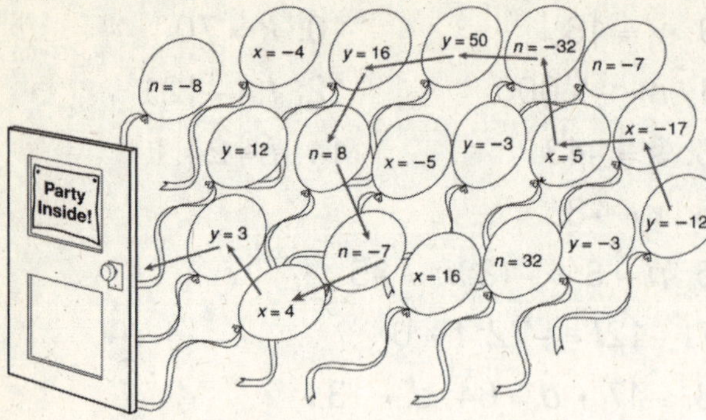

Problem Solving

1. 450 mi
2. \$254 billion
3. 70º
4. Possible answer: 122 years old (in 2008)
5. A
6. G
7. D
8. G

Reading Strategies

1. subtraction
2. addition
3. $w - 12 + 12 = -4 + 12$
4. $w = 8$
5. addition
6. subtraction

Puzzles, Twisters & Teasers

Q: 100
A: 60
U: –63
C: 23
F: –41
K: 17
X: –11
E: –58
O: –8
R: –10
B: 20
S: –6
W: –60
D: –7
G: –9

A BOX OF QUACKERS

LESSON 2-6

Practice A

1. Composite
2. Composite
3. Composite
4. Prime
5. A
6. H
7. B
8. F
9. 5^2
10. $2^2 \cdot 5$
11. 2^5
12. 7^2
13. $2 \cdot 3 \cdot 5$
14. $2^2 \cdot 11$
15. $3^2 \cdot 7$
16. $2^2 \cdot 7$
17. $3 \cdot 31$
18. $2 \cdot 5 \cdot 11$
19. $2^4 \cdot 5$
20. $2 \cdot 3 \cdot 7$
21. 5^4
22. $2 \cdot 7^2$
23. $2^2 \cdot 5 \cdot 7$
24. $2^2 \cdot 23$

Practice B

1. Composite
2. Prime
3. Composite
4. Prime
5. Composite
6. Composite
7. Prime
8. Composite
9. $3 \cdot 19$
10. 7^2
11. $2^3 \cdot 11$
12. $5 \cdot 19$
13. $3 \cdot 5 \cdot 7$
14. $2 \cdot 7^2$
15. $2^2 \cdot 13$
16. $2 \cdot 3 \cdot 7$
17. $2^2 \cdot 17$
18. $7 \cdot 13$
19. $2^2 \cdot 3 \cdot 5$
20. $2^3 \cdot 3^2$
21. $2^3 \cdot 7$
22. $2^4 \cdot 3^2$
23. $2 \cdot 5 \cdot 37$
24. $2^3 \cdot 3 \cdot 7$
25. $2^2 \cdot 31$
26. $5 \cdot 103$
27. $5^2 \cdot 29$
28. $2^2 \cdot 5 \cdot 11$
29. $2 \cdot 3^2 \cdot 5^2$
30. $2^3 \cdot 5^3$
31. $2^4 \cdot 5 \cdot 13$
32. $2^2 \cdot 5^4$
33. 495

Practice C

1. Composite
2. Composite
3. Prime
4. Prime
5. $2^2 \cdot 3 \cdot 7$
6. 7^3
7. $3^3 \cdot 5$
8. $2^2 \cdot 3^2 \cdot 5$

9. $2 \cdot 5 \cdot 29$
10. $2^2 \cdot 3^2 \cdot 5^2$
11. $5^2 \cdot 23$
12. $2 \cdot 3^2 \cdot 5$
13. $2^2 \cdot 5 \cdot 11$
14. $2 \cdot 5^2 \cdot 7$
15. $3^2 \cdot 5 \cdot 11$
16. $2^5 \cdot 3 \cdot 5$
17. $2^4 \cdot 5^3$
18. $2^4 \cdot 5 \cdot 13$
19. $2 \cdot 3 \cdot 5^2 \cdot 11$
20. $2^3 \cdot 3 \cdot 5^3$
21. 972
22. $5 \cdot 30$; $3 \cdot 50$; $6 \cdot 25$; $2 \cdot 75$
23. 13
24. 800
25. 1,485
26. 5,824
27. 62,208
28. 667

Review for Mastery

1. $2^2 \cdot 5$
2. $2^2 \cdot 7$
3. $2 \cdot 3^3$
4. $3^2 \cdot 7$
5. $2^4 \cdot 3$
6. $2^3 \cdot 3$
7. $2^3 \cdot 5$
8. $2 \cdot 7^2$

Challenge

Possible answers are given.

1. 

$3^2 \cdot 5^2$

2. 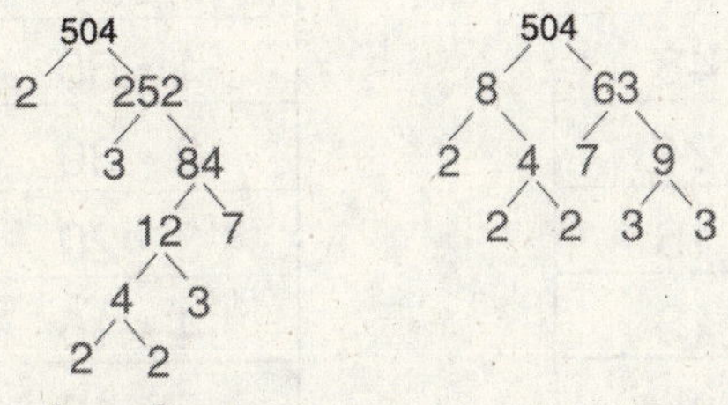

$2^3 \cdot 3^2 \cdot 7$

3. Possible answer: 12 = 7 + 5; 32 = 19 + 13; 44 = 13 + 31
4. Possible answer: 13 = 3 + 5 + 5; 15 = 5 + 5 + 5; 33 = 23 + 5 + 5

Problem Solving

1. 17 ft by 24 ft
2. 19 ft by 18 ft
3. 5 groups of 50; 2 groups of 125; 10 groups of 25
4. 29 pages per day for 15 days
5. C
6. H
7. A
8. J

Reading Strategies

1. A prime number has only two factors, itself and 1. A composite number has more than two factors.
2. $2 \cdot 2 \cdot 2 \cdot 2 \cdot 3$, or $2^4 \cdot 3$
3. $2 \cdot 2 \cdot 2 \cdot 7$, or $2^3 \cdot 7$

Puzzles, Twisters & Teasers

P R I M E F U N I Q U E P
C W E S A H D C B N U O F
T Y P L M A S X D S R W K
E R T Y U J C F T S Y C B
F A C T O R I Z A T I O N
A S Z R A A Q W R E I M F
C G B E X T D S A P B P M
T N M E A W N J I L K O C
O W E R T A T H N B V S I
R Y U I O L C V B N M I L
A S E D I A G R A M F T W
W S C E X P O N E N T E J

PRIME

LESSON 2-7

Practice A

1. B
2. D
3. A
4. F
5. C
6. E
7. J
8. G
9. H
10. K
11. P
12. R
13. N
14. M
15. U
16. Q
17. V
18. W
19. T
20. S
21. 7 bracelets
22. 12 flower baskets

Practice B

1. 3
2. 11
3. 9
4. 5

5. 9
6. 6
7. 4
8. 7
9. 3
10. 2
11. 8
12. 14
13. 25
14. 13
15. 4
16. 7
17. 28
18. 7
19. 15
20. 2
21. 6
22. 8 backpacks
23. 12 baskets

Practice C

1. 15
2. 8
3. 2
4. 20
5. 16
6. 8
7. 5
8. 32
9. 7
10. 91 and 117 have a GCF of 13.
11. 80 and 114 have a GCF of 2.
12. 68 and 187 have a GCF of 17.
13. 9 gift sets
14. 15 frames
15. 24 necklaces

Review for Mastery

1. 3; 9
   3; 6; 8; 12
   3
2. 6; 9; 18
   6; 9; 18; 27
   18
3. 3; 5; 9; 15
   5; 11
   5
4. 2; 3; 4; 6
   2; 4; 5; 10
   4
5. 2; 3; 6
   3; 4; 6
   3; 4; 6; 8
   6
6. 2; 3; 6
   2; 3; 6; 9
   3; 9
   3
7. 9
8. 2
9. 6
10. 5

Challenge

1. planetarium
2. mini-buses
3. sandwich and juice

On Friday, your child will be taking a school trip to the planetarium. The school will pay $60 for the seventh-graders' tickets and $90 for the eighth-graders' tickets. The PTA has graciously offered to pay $175 for 5 mini-buses to transport the students. Each class's student council will pay for its class's lunches. Both classes will buy 10 sandwich and juice packs for a total cost of $60 for the seventh graders and $90 for the eighth graders.

Problem Solving

1. 15 inches
2. 24 sandwiches
3. 40 miles per day
4. 6 servings
5. D
6. H
7. B
8. J

Reading Strategies

1. 1, 18, 2, 9, 3,6
2. 1, 2, 3, 6, 9, 18
3. 1, 21, 7, 3
4. 1, 3, 7, 21
5. 3
6.

48
1 • 48
2 • 24
3 • 16
4 • 12
6 • 8

7.

60
1 • 60
2 • 30
3 • 20
4 • 15
5 • 12
6 • 10

8. 1, 2, 3, 4, 6, 12
9. 12

Puzzles, Twisters & Teasers

1. 12
2. 15
3. 10
4. 2
5. 3
6. 7
7. 1

8. 9

A CORN KERNEL

LESSON 2-8

Practice A

1. 21
2. 20
3. 60
4. 75
5. 33
6. 36
7. 42
8. 40
9. 78
10. 90
11. 60
12. 30
13. 56
14. 36
15. 70
16. 90
17. 60
18. 36
19. 32
20. 60
21. 126
22. 15th beat
23. 24 minutes

Practice B

1. 40
2. 30
3. 18
4. 48
5. 90
6. 55
7. 45
8. 28
9. 28
10. 60
11. 180
12. 15
13. 98
14. 96
15. 150
16. 90
17. 60
18. 36
19. 72
20. 240
21. 90
22. 75 minutes
23. 60 minutes
24. 24 minutes

Practice C

1. 36
2. 180
3. 45
4. 100
5. 77
6. 48
7. 21
8. 200
9. 90
10. 78
11. 120
12. 200
13. 72
14. 36
15. 72
16. 60
17. 100
18. 56
19. 60
20. 126
21. 320
22. 24 minutes
23. 75 minutes
24. 84 minutes

Review for Mastery

1. 36
2. 21; 28; 35; 42; 49; 56
   18; 24; 30; 36; 42
   42
3. 20; 40; 60; 80; 100
   15; 30; 45; 60
   60
4. 36
5. 80
6. 36

Challenge

1. 6; 12
   12
2. 2; 10
   10
3. 2; 3
   6
4. 2; 10
   10
5. 4; 6
   12
6. 10; 15
   30
7. 5; 10
   10
8. 14; 35
   70
9. 10; 15
   30
10. 3; 12
   12
11.

T	H	E		C	H	E	E	T	A	H
30	10	12		70	10	12	12	30	6	10

Problem Solving

1. 2014
2. 2012
3. 10 minutes
4. 15,000 mi; 30,000 mi
5. D
6. H
7. D
8. G

Reading Strategies

1. Possible answer; the answers or products when you multiply by a certain number
2. something that is shared
3. multiples that two or more numbers share, or have in common
4. the smallest multiple that two or more numbers have in common
5. 2, 4, 6, 8, 10, 12, 14, 16, 18
6. 3, 6, 9, 12, 15, 18, 21, 24, 27
7. 6, 12, and 18
8. 6

Puzzles, Twisters & Teasers

1. 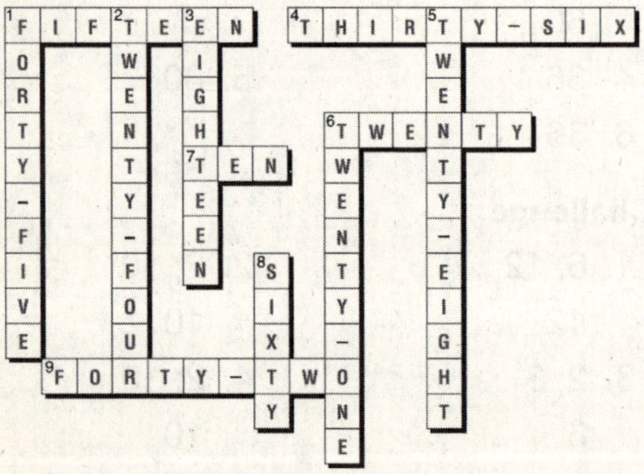

LESSON 2-9

Practice A

Possible answers given.

1. $\frac{2}{8}$
2. $\frac{6}{10}$
3. $\frac{8}{18}$
4. $\frac{2}{16}$
5. $\frac{2}{12}$
6. $\frac{1}{2}$
7. $\frac{2}{5}$
8. $\frac{7}{10}$
9. $\frac{6}{9}, \frac{6}{9}$; yes
10. $\frac{21}{30}, \frac{16}{30}$; no
11. $\frac{12}{24}, \frac{12}{24}$; yes
12. $\frac{80}{40}, \frac{80}{40}$; yes
13. $\frac{18}{12}, \frac{18}{12}$; yes
14. $\frac{16}{24}, \frac{15}{24}$; no
15. $1\frac{1}{4}$
16. $4\frac{1}{2}$
17. $2\frac{3}{5}$
18. $1\frac{5}{7}$
19. $\frac{11}{3}$
20. $\frac{19}{8}$
21. $\frac{8}{5}$
22. $\frac{17}{6}$
23. $\frac{23}{2}$
24. no

Practice B

Possible answers given.

1. $\frac{4}{18}$
2. $\frac{16}{30}$
3. $\frac{14}{16}$
4. $\frac{2}{3}$
5. $\frac{3}{5}$
6. $\frac{3}{4}$

Possible answers given.

7. $\frac{24}{30}, \frac{24}{30}$; yes
8. $\frac{18}{24}, \frac{16}{24}$; no
9. $\frac{24}{72}, \frac{36}{72}$; no
10. $\frac{35}{20}, \frac{36}{20}$; no
11. $\frac{60}{48}, \frac{60}{48}$; yes
12. $\frac{30}{18}, \frac{30}{18}$; yes
13. $2\frac{5}{8}$
14. $9\frac{1}{4}$
15. $3\frac{1}{5}$
16. $5\frac{4}{9}$
17. $\frac{26}{3}$
18. $\frac{19}{12}$
19. $\frac{103}{4}$
20. $\frac{47}{6}$
21. $\frac{135}{4}$
22. yes

Practice C

Possible answers given.

1. $\frac{14}{24}, \frac{21}{36}$
2. $\frac{22}{30}, \frac{33}{45}$
3. $\frac{28}{50}, \frac{42}{75}$
4. $\frac{9}{10}, \frac{36}{40}$
5. $\frac{5}{1}, \frac{10}{2}$
6. $\frac{9}{8}, \frac{18}{16}$

Possible answers given.

7. $\frac{40}{60}, \frac{36}{60}$; no
8. $\frac{20}{36}, \frac{27}{36}$; no
9. $\frac{36}{48}, \frac{36}{48}$; yes
10. $\frac{35}{40}, \frac{36}{40}$; no
11. $\frac{96}{18}, \frac{96}{18}$; yes
12. $\frac{76}{30}, \frac{75}{30}$; no
13. $6\frac{1}{7}$
14. $5\frac{2}{5}$
15. $6\frac{7}{8}$
16. $7\frac{3}{11}$
17. $\frac{127}{5}$
18. $\frac{106}{9}$
19. $\frac{59}{12}$
20. $\frac{85}{16}$
21. $\frac{139}{12}, 11\frac{7}{12}$
22. 210 quarter cups

Review for Mastery

Possible answers given.

1. $\frac{2}{4}$
2. $\frac{2}{8}$
3. $\frac{4}{10}$
4. $\frac{4}{6}$
5. 54; 60 no
6. 90; 90; yes
7. 24; 27; no
8. 4; 4; 4; 1; $3\frac{1}{4}$
9. 3; $\frac{2}{5}$; 15; $\frac{2}{5}$; $\frac{17}{5}$
10. $4\frac{3}{5}$

11. $1\frac{7}{8}$
12. $7\frac{5}{7}$
13. $4\frac{2}{3}$
14. $\frac{11}{2}$
15. $\frac{27}{8}$
16. $\frac{41}{15}$
17. $\frac{35}{4}$

Challenge

1. $5\frac{1}{3}$
2. $\frac{7}{10}$
3. $1\frac{1}{2}$
4. $\frac{12}{20}$
5. $\frac{23}{8}$
6. $2\frac{4}{5}$
7. $\frac{8}{18}$
8. $\frac{2}{3}$
9. $\frac{6}{27}$
10. $6\frac{2}{3}$
11. $\frac{15}{45}$
12. $\frac{19}{6}$
13. $\frac{1}{4}$
14. $\frac{54}{10}$
15. $3\frac{3}{4}$
16. $\frac{35}{70}$
17. $\frac{42}{5}$

Problem Solving

1. $19\frac{3}{7}$ miles
2. Possible answer: $\frac{9}{2}; \frac{18}{4}$
3. $\frac{52}{4}$ inches
4. $4\frac{5}{6}$ feet
5. A
6. J
7. C
8. G

Reading Strategies

1. yes
2. Possible answer: You can multiply both the numerator and denominator by 6 to get $\frac{6}{12}$.
3. no
4. Possible answer: You cannot multiply both 1 and 5 by the same number to get 2 and 3.
5. no
6. Possible answer: You cannot divide both 7 and 15 by the same number to get 1 and 7.
7. yes
8. Possible answer: You can divide both 3 and 9 by 3 to get $\frac{1}{3}$.

Puzzles, Twisters & Teasers

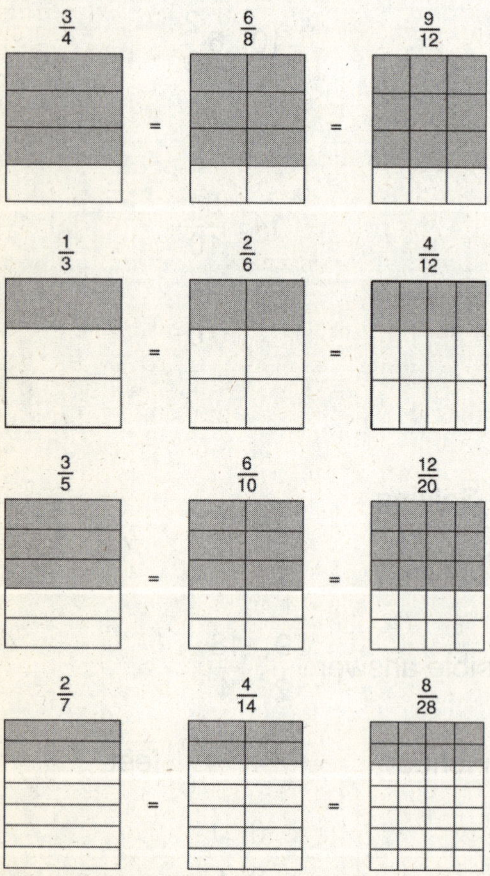

IT SAW THE SALAD DRESSING

LESSON 2-10

Practice A

1. 0.7
2. 0.45
3. 0.75
4. 0.8
5. 0.375
6. 1.4
7. 3.0
8. 1.67
9. 0.3
10. 0.8
11. 0.04
12. 0.15
13. $\frac{11}{20}$
14. $\frac{3}{100}$
15. $-\frac{3}{4}$
16. $2\frac{1}{10}$ or $\frac{21}{10}$
17. $5\frac{1}{4}$ or $\frac{21}{4}$
18. $9\frac{33}{100}$ or $\frac{933}{100}$
19. $\frac{9}{5}$ or $1\frac{4}{5}$
20. $-\frac{87}{50}$ or $-1\frac{37}{50}$
21. $\frac{53}{5}$ or $10\frac{3}{5}$
22. $-7\frac{2}{25}$ or $-\frac{177}{25}$
23. $\frac{5}{8}$
24. $\frac{1}{1{,}000}$
25. 0.422
26. 0.867

Practice B

1. 0.2
2. 0.95
3. 0.625
4. 2.2
5. 3.167
6. 4.25
7. 1.083
8. 4.286
9. 1.75
10. 0.45
11. 1.1
12. 0.08
13. $\frac{17}{20}$
14. $\frac{11}{100}$
15. $-\frac{1}{4}$
16. $4\frac{3}{10}$ or $\frac{43}{10}$
17. $7\frac{3}{4}$ or $\frac{31}{4}$
18. $5\frac{3}{100}$ or $\frac{503}{100}$
19. $-1\frac{3}{50}$ or $-\frac{53}{50}$
20. $\frac{3}{8}$

21. $-2\frac{13}{20}$ or $-\frac{53}{20}$

22. $-5\frac{3}{5}$ or $-\frac{28}{5}$

23. $1\frac{3}{25}$ or $\frac{28}{25}$

24. $\frac{1}{200}$

25. 0.063

26. 0.853

Practice C

1. 0.417
2. 4.75
3. 3.571
4. 1.125
5. 5.667
6. 1.182
7. 1.667
8. 2.308
9. 0.18
10. 0.52
11. 1.15
12. 4.25
13. $\frac{1}{4}$
14. $\frac{13}{100}$
15. $-\frac{13}{20}$
16. $4\frac{4}{5}$ or $\frac{24}{5}$
17. $1\frac{7}{8}$ or $\frac{15}{8}$
18. $9\frac{3}{50}$ or $\frac{453}{50}$
19. $-1\frac{1}{125}$ or $-\frac{126}{125}$
20. $\frac{5}{8}$
21. $-2\frac{1}{20}$ or $-\frac{41}{20}$
22. no
23. yes
24. no
25. 0.120
26. $\frac{4{,}256}{14{,}053}$, 0.303

Review for Mastery

1. 0.4; 0.4
   2 0
   0
2. 0.75
3. 0.875
4. 1.5
5. 1.667
6. $\frac{4}{5}$
7. $\frac{9}{4}$ or $2\frac{1}{4}$
8. $-\frac{1}{50}$

Challenge

1. r
2. n
3. r
4. n
5. n
6. n
7. r
8. r
9. n
10. n
11. r
12. n
13. 0.01020304…
14. 8.63633633363333…
15. 25.12123123412345…
16. 9.98979695…
17. Possible answer: 1.2222…
18. Possible answer: 1.191991999…

Problem Solving

1. 0.925
2. 0.103
3. 1.4 in.
4. 0.797
5. D
6. H
7. C
8. J

Reading Strategies

1. No. Possible answer: When you subtract 20 from 20 there is a remainder of 0.
2. a terminating decimal
3. Possible answer: There is a remainder when dividing 2 by 6, but no remainder when dividing 2 by 5.
4. repeating decimal

Puzzles, Twisters & Teasers

D	E	N	O	M	I	N	A	T	O	R	U	J	T	K	P
I	E	R	S	T	A	T	I	S	T	I	C	F	E	T	A
V	W	C	A	V	E	R	A	G	E	N	O	D	R	K	T
I	F	V	I	Z	X	C	V	B	Y	H	N	U	M	W	T
D	R	Q	E	M	T	G	B	Y	H	A	V	L	I	Z	E
E	A	S	D	U	A	B	G	R	V	I	E	C	N	I	R
P	C	O	L	C	O	L	P	I	C	K	R	O	A	D	N
N	U	M	E	R	A	T	O	R	A	U	T	G	T	E	I
I	I	W	E	R	P	A	R	E	P	E	A	T	I	N	G
U	O	Z	A	Q	W	S	X	C	O	T	G	H	N	J	I
Y	N	G	F	R	A	C	T	I	O	N	A	T	G	O	R

I C D K (I see decay.)

LESSON 2-11

Practice A

1. <
2. <
3. >
4. <
5. <
6. >
7. <
8. >
9. >
10. <
11. <
12. <
13. $0.5, 0.\overline{5}, \frac{5}{8}$
14. $1.3, 1\frac{1}{3}, 1.34$
15. $2.07, 2.67, 2\frac{7}{10}$
16. $-4.8, -4\frac{1}{8}, -4.08$
17. $-5.9, -5\frac{7}{8}, 5.78$
18. $0.8, \frac{5}{6}, \frac{6}{7}$
19. $2.65, 2\frac{2}{3}, 2.69$
20. $-0.38, -\frac{3}{8}, -0.04$
21. $\frac{3}{7}, \frac{4}{9}, 0.45$
22. $\frac{7}{12}, 0.583$
23. Joella

Practice B

1. <
2. <
3. <
4. >
5. <
6. >
7. <
8. >
9. <
10. >
11. <
12. <
13. $0.4, \frac{4}{9}, 0.45$
14. $1.65, 1\frac{2}{3}, 1.7$
15. $3\frac{1}{8}, 3.18, 3.80$
16. $-5\frac{2}{5}, -5.25, -5$
17. $-6\frac{3}{4}, -6.4, 6.34$
18. $\frac{8}{9}, 0.91, \frac{11}{12}$
19. $-\frac{5}{7}, -0.65, -\frac{3}{5}$
20. $0.3, \frac{1}{3}, 0.345$
21. $-0.75, -\frac{5}{8}, \frac{7}{8}$
22. $\frac{373}{500}$; 0.746
23. Mercury

Practice C

1. >
2. <
3. >
4. <
5. <
6. <
7. <
8. >
9. >
10. <
11. <
12. <
13. $0.54, \frac{9}{16}, 0.65$
14. $3.14, 3\frac{4}{9}, 3.49$
15. $-3.5, -3\frac{1}{10}, -3.05$
16. $-2.7, -2\frac{7}{12}, 2.59$
17. $\frac{5}{6}$
18. 0.095
19. −0.025
20. −5

21. $\frac{525}{746}$, 0.704

22. Jupiter

Review for Mastery

1. <
2. >
3. <
4. >
5. >
6. <
7. >
8. <

Challenge

Possible answers given.

1. $\frac{1}{10}$
2. 3.41
3. 1.1
4. –7.2
5. –12.15
6. $-\frac{1}{4}$
7. $1\frac{1}{4}$
8. $-2\frac{1}{5}$
9. 3.015
10. 4.3
11. –1.01
12. 3.6
13. –2.5
14. –0.1
15. $7\frac{1}{4}$
16. $5\frac{9}{10}$
17. $-\frac{7}{10}$
18. 3.195

Problem Solving

1. $14\frac{1}{4}$, 14.69, $14\frac{7}{10}$, $14\frac{4}{5}$
2. 2001: 2.9; 2002: $2\frac{3}{4}$;

   2003: $2\frac{2}{5}$
3. 0.304, 0.334, 0.343
4. George
5. C
6. J

Reading Strategies

1. >
2. <
3. <
4. =
5. =
6. >

Puzzles, Twisters & Teasers

1. O
2. T
3. H
4. E
5. M
6. A
7. T
8. I
9. C
10. S

M O T H E M A T I C S